パイロットのための
ICAO航空英語能力試験教本

Simon Cookson & Michael Kelly

CD-ROM付

READY FOR DEPARTURE !

成山堂書店

本書の内容の一部あるいは全部を無断で電子化を含む複写複製（コピー）及び他書への転載は，法律で認められた場合を除いて著作権者及び出版社の権利の侵害となります。成山堂書店は著作権者から上記に係る権利の管理について委託を受けていますので，その場合はあらかじめ成山堂書店（03-3357-5861）に許諾を求めてください。なお，代行業者等の第三者による電子データ化及び電子書籍化は，いかなる場合も認められません。

FOREWORD

本書は、ICAO（国際民間航空機関）が打ち出した航空機事故防止策の一つとして、ほぼ10年近く前に180カ国以上からなる加盟国の総会で可決した、航空操縦士（管制官も含む）の英語能力測定テストのための準備教材として著されたものである。

このテストは、当時、航空機事故の中にパイロットと管制官の間のコミュニケーション不足が原因で起きたものがある、との指摘から、特に国際線を飛ぶパイロットの英語によるコミュニケーション能力を保証する必要があるとされた。もちろん、パイロットは、誰もが通常の航空英語に関しては問題なく使いこなすことができるが、いざ、異常事態が起こった時に、いつも使っている決まり文句としての航空英語では不十分になってくる。そこで、そのような緊急事態が起きた時の英語能力がどれだけあるかを測る必要があるのである。例えば、機内で火災が起こったり、急病人が出たり、あるいは、また、悪天候など、予想外の出来事で機体に何らかの損傷があったりした場合、それを管制官、あるいは他のパイロットにどうやって伝えるのか。このような緊急事態の場合、決まり文句の航空英語では不十分で、どうしても日常的な英語（Plain English）が使えなければならないのである。

通常の航空英語は、パターン化され、100%分かる発音でなくても、聞き手は容易に内容を予測できる。しかし、緊急事態が起こった場合は、その状況を説明するための英語自体が分かりやすくなければならない。つまり、緊急事態の場合は、それを説明する英語がまずければ、コミュニケーションに支障をきたしかねないのである。

2005年から上智大学の国際言語情報研究所では、ICAOの基準に基づいた航空英語能力テストの開発を行った。本書の著者の一人でもあるKelly氏は、当時、JALで英語の研修を担当していたが、本テスト開発に非常に熱心に取り組まれた。

本書の内容は、正に航空英語能力テストの内容を忠実に踏まえて作られており、今後、国際線を飛ぶパイロットの英語力の育成に大いに役立つものと信じる。日本人は英語が弱い、とよく言われるが、国際線を飛ぶパイロットは、国境を越え、世界中の人々とコミュニケーションできなければならない。その意味で、本教材は単に航空英語能力テストのための準備教材、という狭い位置づけではなく、様々な状況で英語を使う訓練をすることで、より一般的な英語力の育成につながることを期待したい。

吉田研作
上智大学教授
国土交通省航空局航空英語能力証明審査会会長

2012年4月

ACKNOWLEDGMENTS

This book is dedicated to President Toyoshi Satow, Chancellor of J. F. Oberlin University and Affiliated Schools, who had the vision and determination to establish the Flight Operations Program at J. F. Oberlin University.

A lot of people have contributed to the book, and we would like to give particular thanks to the following:
- Professor Kensuke Yoshida of Sophia University for checking the manuscript and for writing the foreword;
- Professor Rocco Sorrenti of J. F. Oberlin University for his unstinting positivity and support, as well as his help with the artwork and audio recordings;
- Assistant Professor Chihiro Tajima of Rikkyo University for checking the manuscript, advising on material design, and writing the Japanese sections of the book;
- Marina Kuribayashi, Ayumi Inoue and Suvd Jargalsaikhan for contributing all the pictures in the book;
- and Benedict Rowlett, Peter Mattersdorf, Ryota Isono, Shoji Ota, Ryota Ishitobi and Keitaro Kobayashi for recording the audio tracks and Shu Wada for hunting down mistakes.

We would also like to thank the following members of J. F. Oberlin University for their help and support:
- Captain Kunio Miyazaki, Captain Mitsuru Okada, Professor Yoshimasa Suzuki, Professor Takaharu Ishihara, Professor Takakazu Haraguchi, Professor Saburo Onodera, Professor Bruce Batten, Professor Reiko Kobayashi, Tadao Hamada, Shin Kitamura, Akiko Inada, Takahide Sakamoto, Tim Marchand, Robert Russell, Mihoko Inamori, and all the students of the Flight Operations Program.

While a lot of people have contributed to the book, we would like to make clear that any mistakes that remain in the manuscript are the sole responsibility of the authors.

Finally, this book could not have been published without the guidance and support of the staff of Seizando-Shoten Publishing Co. We are especially grateful to Yosuke Itagaki, Takashi Ono and the company president, Noriko Ogawa, for all their patience!

To our readers, we wish you happy and safe flying!

Simon Cookson and Michael Kelly, January 2013

Flight Operations Program
Aviation Management Department
J. F. Oberlin University
Tokyo, Japan

CONTENTS

Section	Pages	Content of Section	Flight Phase
Introduction	2-9	The ICAO English Language Proficiency Test	
Unit 1	10-15	"Ladies And Gentlemen…" Delays Due To Weather	Pre-Flight Operations
Unit 2	16-21	"Use Caution For Icy Conditions" Snow & Ice At The Airport	At The Ramp
Unit 3	22-27	"Hold Your Position!" Obstructions On The Taxiway	Ground Movement
Unit 4	28-33	"Stand By For Clearance" Rejecting Takeoff	Cleared For Takeoff
Unit 5	34-39	"Check And Confirm" Air Turnback	Takeoff & Climb
Unit 6	40-45	"Level Off At Your Discretion" Equipment Failure & Rough Air	Climb
Review 1	46-49	Review of Units 1-6	
Unit 7	50-55	"Is There A Doctor On Board?" Passenger Injuries & Problems	Cruise
Unit 8	56-61	"Mayday, Mayday, Mayday!" Smoke In The Cabin	Emergency
Unit 9	62-67	"The Airport Is Now Closed" Bad Weather & Natural Disasters	Holding
Unit 10	68-73	"Down And Locked?" Problems During Approach	Approach
Unit 11	74-79	"Go Around, Go Around" Crosswinds & Wake Turbulence	Landing
Unit 12	80-85	"Request Emergency Assistance" Overruns & Other Mishaps	After Landing
Review 2	86-89	Review of Units 7-12	
Answer Key	90-105	Answers & Transcripts	
Self-Evaluation	106	Checklist for the 6 ICAO Areas Of Evaluation	

INTRODUCTION

THE NEED FOR ENGLISH TESTING

Aviation accident rates are much lower than they used to be, but safety experts are constantly trying to find ways of improving safety. In recent years mechanical failures have not been so prominent in aircraft accidents, so more attention has been given to human factors that contribute to accidents. One area of human factors which has received attention is communication, especially the English language proficiency of operators. Safety experts have identified three ways in which language may be involved in aviation accidents and incidents:

1. the incorrect use of ICAO standard phraseology;
2. lack of plain language proficiency;
3. the use of more than one language in the same airspace.

In recognition of the importance of English language proficiency, the International Civil Aviation Organization (ICAO) has established a licensing regulation that requires all international pilots and air traffic controllers to demonstrate proficiency in the English language. This regulation became Japanese law on March 5th 2008 and it became an ICAO regulation on March 5th 2011.

Each country is responsible for making its own English language proficiency test, but the test must be based on the ICAO approved rating scale and the areas of evaluation outlined in the *Manual on the Implementation of ICAO Language Proficiency Requirements* (DOC 9835). The following pages have information about the English language proficiency test that was developed in Japan by the Japanese Civil Aviation Bureau (JCAB).

ICAO ENGLISH LANGUAGE PROFICIENCY TEST

FREQUENTLY ASKED QUESTIONS

- *Who needs to take the test?*
  All pilots who operate in international airspace need a JCAB-issued English language proficiency license. If you do not fly in international airspace, then you do not need this license, but you will be restricted to domestic flights in Japan.

- *What kind of test is it?*
  This test has the specific purpose of testing your proficiency at understanding and communicating in aviation English. The test has a listening comprehension section and an interview section. All the questions and tasks are aviation-related, but it is not a knowledge test. If you do not know an answer to a question, then you should clearly say so.

- *What do I have to do?*
  You have to demonstrate listening and speaking skills. The focus is on your ability to use both ICAO standard phraseology and plain English. Note that the test does not have a reading or writing section.

- *Where can I get more information?*
  The JCAB website has more information. Access the website at http://www.mlit.go.jp/index.html and click on the「国土交通省について」and then the「国家試験のご案内」links. The「航空」section has information about the test, including questions from the listening section of previous tests.

INTRODUCTION

英語能力テストの必要性

航空事故の起きる率は、以前に比べて低くなっているが、航空安全の専門家は、安全性を向上させる方法を常に模索している。近年、メカニカルな問題が関わる航空事故は目立たなくなっているために、ヒューマンファクターと呼ばれる要因が事故を引き起こすことに注目が集まっている。ヒューマンファクター要因の中で最も注目されている領域の一つは、コミュニケーションで、特に航空従事者の英語能力が重要視されている。航空安全の専門家は、言語が関わる航空事故に対し、以下の3つの要素を挙げている：

1. ICAO の「Standard Phraseology」の誤使用；
2. 不十分な日常英語能力；
3. 同空域での、複数言語の使用。

英語能力の重要性が認知されるとともに、ICAO: The International Civil Aviation Organization（国際民間航空機関）は、国際線パイロットと管制官に対し、一定の英語能力を求める証明に関する規定を定めた。この規定は、2008年3月5日に、日本の法律でも定められ、2011年3月5日には、ICAO規定によっても、正式に定められたのである。

英語能力試験の作成については、各国にその責任が委ねられているが、試験では ICAO の「Manual on the Implementation of ICAO Language Proficiency Requirements (DOC 9835)」に定められた評価項目と評価基準に基づいていなければならない。以下に JCAB：the Japanese Civil Aviation Bureau（国土交通省航空局）が開発した英語能力試験に関する情報を掲載する。

ICAO 航空英語能力試験

よくある質問

- 受験が必要なのは誰ですか。

  国際空域で作業する全てのパイロットは、JCAB により発行された航空英語能力証明が必要です。国際空域を飛ばなければ、この証明は必要ありませんが、その場合日本の国内線に限られてしまいます。

- どのような試験ですか。

  試験の目的は、航空英語コミュニケーションへの理解度を測ることであり、リスニングセクションと、インタビューセクションがあります。全ての問題やタスクは航空に関連しますが、知識を問う試験ではありません。したがって、試験問題が理解できない場合には、その旨を発言し伝えなければなりません。

- 何を求められているのでしょうか。

  リスニングとスピーキング技能を示さなければなりません。焦点は、ICAO の「Standard Phraseology」、および日常英語が使用できるかです。リーディングとライティングのセクションはありません。

- さらに情報を得るにはどうすればよいですか。

  JCAB のサイトに詳しい情報が載っています。まずは、http://www.mlit.go.jp/index.html サイトにアクセスし、そこから「国土交通省について」、次に「国家試験のご案内」、その後「航空」とクリックしていきます。このページには、過去問の情報のリンクも載っていますから、リンクをクリックすれば、過去問を使ったリスニングセクションの勉強も可能です。

INTRODUCTION

ICAO LANGUAGE PROFICIENCY LEVELS

You will be evaluated based on the 6 levels of English language proficiency shown in the table below. The minimum operational level for international pilots and air traffic controllers is Level 4. The license for a Level 4 rating is valid for 3 years. Level 5 is valid for 6 years, and Level 6 (or Expert Level) is indefinite.

LEVEL	NAME	PASS/FAIL	VALID FOR
Level 6	Expert Level	Pass	Indefinite
Level 5	Extended Level	Pass	6 years
Level 4	Operational Level	Pass	3 years
Level 3	Pre-Operational Level	Fail	Not valid
Level 2	Elementary Level	Fail	Not valid
Level 1	Pre-Elementary Level	Fail	Not valid

ICAO COMMUNICATION TASKS

ICAO has outlined 5 communication tasks that pilots must demonstrate in order to show their English language proficiency. Pilots must be able to:

1. communicate effectively by radio and in face-to-face situations;
2. communicate on work-related topics with accuracy and clarity;
3. exchange messages and recognize and resolve misunderstandings;
4. handle successfully a complication or unexpected turn of events;
5. use a dialect or accent which is intelligible to a proficient listener.

Communication skill is essential for flight safety!

All 5 descriptors are covered in the listening and interview sections of the test. The focus of the test is evaluating the English language proficiency of pilots in their working environment, and you will only be tested on your listening and speaking ability. Below is information about each section of the test.

TEST SECTION 1: LISTENING

The listening section of the test has 14 ATC dialogs. You listen to a short ATC dialog, then answer 3 questions about the dialog. There are 42 questions in total. To pass the listening section you must answer 70% or more of the questions correctly. Note that you must pass the listening section before you can take the interview section. Here is a sample dialog and question.

SAMPLE DIALOG

CONTROLLER:	OB 1663, snow removal now in progress on all runways, expect delay of 30 minutes or more.
PILOT:	Chitose Ground, OB 1663, roger, request taxi back to ramp area for de-ice.
CONTROLLER:	OB 1663, taxi back to ramp at your discretion, use extreme caution, ground visibility limited due to snow flurries.
QUESTION	1. OB 1663 will…
	a. hold his position for 30 minutes.
	b. taxi to the runway with caution.
	c. do de-ice procedure at the ramp.
	d. request pushback in 30 minutes.

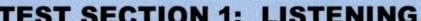

INTRODUCTION

ICAO の航空英語能力レベル

航空英語能力は、以下の表が示すように、6レベルで評価されます。国際線パイロットと管制官に必要最低限なのは、レベル4（Operational Level）です。レベル4の証明は3年間有効です。レベル5は6年間、そしてレベル6（Expert Level）は永久に有効となります。

レベル	名称	合格／不合格	有効期限
Level 6	Expert Level	合格	永久
Level 5	Extended Level	合格	6年
Level 4	Operational Level	合格	3年
Level 3	Pre-Operational Level	不合格	国際線不可
Level 2	Elementary Level	不合格	国際線不可
Level 1	Pre-Elementary Level	不合格	国際線不可

ICAO 航空英語の課題

ICAO は、以下の5つの課題を示しており、パイロットはそれぞれを行い英語能力を示さなければなりません。パイロットは：

1. 無線、および対面の状況において、効果的にコミュニケーションを行う；
2. 仕事に関わる話題について、正確かつ明確にコミュニケーションを行う；
3. メッセージをやり取りし、誤解に気づき、解決する；
4. 複雑な事や想定外の出来事を、うまく切り抜ける；
5. 聞き手が理解できるような、方言やアクセントを使用する。

5つ全ての課題が、試験のリスニングとインタビューのセクションでカバーされています。試験の焦点は、パイロットが仕事をする環境での英語能力を評価することであり、リスニングとスピーキング技能のみが試験の対象です。以下は試験の各セクションの概要です。

試験セクション1：リスニング

リスニングセクションは、14の ATC 会話から成っています（ATC: Air Traffic Control: 管制官）。短い ATC 会話 を聞き、それぞれの会話に対して3問に答えます。全部で42問から成っています。リスニングセクションをパスするには、70%以上を正解する必要があります。リスニングセクションをパスしなければ、インタビューセクションの受験ができません。以下は ATC 会話と問題の例です。

SAMPLE DIALOG

CONTROLLER: OB 1663, snow removal now in progress on all runways, expect delay of 30 minutes or more.

PILOT: Chitose Ground, OB 1663, roger, request taxi back to ramp area for de-ice.

CONTROLLER: OB 1663, taxi back to ramp at your discretion, use extreme caution, ground visibility limited due to snow flurries.

QUESTION
1. OB 1663 will…
   a. hold his position for 30 minutes.
   b. taxi to the runway with caution.
   c. do de-ice procedure at the ramp.
   d. request pushback in 30 minutes.

INTRODUCTION

TEST SECTION 2: INTERVIEW

The interview section has 3 stages: single picture, ATC description, and picture sequence. Each stage takes about 5 minutes, so the entire interview lasts 15-20 minutes. Each stage serves a different purpose:
1. *Single Picture:* a single picture card is used to describe a situation;
2. *ATC Description:* an ATC role play is acted out;
3. *Picture Sequence:* a picture sequence card is used to describe a sequence of events.

THE 6 AREAS OF EVALUATION

Your interview is recorded and evaluated by a JCAB-designated rater. The rater listens to the interview and evaluates your proficiency in 6 areas. The lowest evaluation of the 6 areas will be your final evaluation. You must receive at least a Level 4 rating in all 6 areas of evaluation in order to get a license. This table shows the areas of evaluation.

AREAS OF EVALUATION	REQUIREMENTS FOR LEVEL 4
Pronunciation	pronunciation is intelligible to the aeronautical community
Structure	structure is used creatively and usually controlled
Vocabulary	the range and accuracy of vocabulary is usually sufficient
Fluency	stretches of language are produced at an appropriate tempo
Comprehension	listening comprehension is mostly accurate on work-related topics
Interactions	responses are usually appropriate and informative

KEY POINTS TO REMEMBER

- *There is no answer key!*
  The test does not have an answer key. Many pilots try to memorize answers for the test, but this does not work well. There are many different picture cards and ATC situations, so interviewers can choose from a large number of combinations. You must demonstrate that you can communicate effectively in any situation.
- *Learn to evaluate your proficiency.*
  Remember the 6 areas of evaluation, and know your level of English proficiency in each area. This will help you understand how you need to improve.
- *Know your voice.*
  Listen to your voice in English. Record yourself and play back the recording. Is your pronunciation clear? Can you understand what you are saying? Are there many pauses or fillers? Are your responses long enough? It is important to understand how you sound to other people. If possible, have other people listen and ask them if they understand your responses.
- *Always check and confirm.*
  Make sure that you understand the task or question. If you are not sure, then check and confirm your understanding.

INTRODUCTION

試験セクション2：インタビュー

インタビューセクションは、3つのステージから成っています：「Single Picture」、「ATC Description」、「Picture Sequence」。各ステージの所要時間は、約5分であるために、インタビュー全体では15〜20分かかることになります。各ステージには、それぞれの目的があります：

1. 「Single Picture」: Single Picture カードは、状況を描写するために用いられる；
2. 「ATC Description」: ATC のロールプレイをする；
3. 「Picture Sequence」: Picture Sequence カードは、事柄の順番の描写に用いられる。

6つの評価項目

インタビューは、録音され JCAB が指定する試験官によって評価されます。試験官は、インタビューを聞いて、6つの評価項目において評価を下します。6つの評価項目で受けた評価のうち、最も低い評価が、最終得点となります。証明を取得するには、6つの評価項目の全てにおいて、レベル4以上の評価を得る必要があります。以下が6つの評価項目です：

6つの評価項目	レベル4に必要なこと
Pronunciation	pronunciation is intelligible to the aeronautical community
Structure	structure is used creatively and usually controlled
Vocabulary	the range and accuracy of vocabulary is usually sufficient
Fluency	stretches of language are produced at an appropriate tempo
Comprehension	listening comprehension is mostly accurate on work-related topics
Interactions	responses are usually appropriate and informative

キーポイント

- 回答集はありません！
  試験には、回答集がありません。試験の答えを暗記しようとするパイロットが多いのですが、これはあまり良く ありません。「Picture Card」には、さまざまな ATC の状況が示されていますし、インタビューの試験官は、数ある組み合わせの中からカードを選びます。したがって、どのような状況でも効果的にコミュニケーションをとれることを示すことが重要なのです。

- 自身の能力を評価することを学びましょう。
  6つの評価項目（6 Areas of Evaluation）を思い出してください。そして、それぞれの項目に対する自身の英語能力のレベルを知りましょう。そうすることで、どのような改善が必要なのか分かるようになります。

- 自身の声を知りましょう。
  自身の英語の声を聞きましょう。録音して聞き返します。発音は明確ですか。自身が何を言っているか理解できますか。沈黙やつなぎ語が多くないですか。回答は十分な長さですか。自身が他者にどのように聞こえているのかを知ることは重要です。可能であれば他者に聞いてもらい、回答が理解できるか尋ねてみてください。

- 常に確かめ確認します。
  指示や問題が理解できていなければなりません。自信がない場合は、試験官に確かめて確認しましょう。

INTRODUCTION

ABOUT THIS BOOK

CONTENTS

This book has 12 units, each of which is structured around the 3 stages of the interview section of the test. Each unit contains 6 pages:

- *Introduction:* introduces key words for the unit;
- *Single Picture:* practices the single picture stage of the interview section;
- *ATC Description:* practices the ATC description stage of the interview section;
- *Useful Language:* introduces useful grammar structures;
- *Picture Sequence:* practices the picture sequence stage of the interview section;
- *Aviation Reading:* introduces extra vocabulary and a reading activity.

All the units provide plenty of practice in the 6 areas of evaluation, as well as useful advice about the different stages of the interview section. Units 1-3 and 7-9 have a special focus on pronunciation, with exercises for the B/V, S/TH and L/R sounds, which are problematic for many Japanese people. Units 4-6 and 10-12 have extra listening comprehension exercises.

HOW TO USE THE BOOK

- *Build your vocabulary range!*

  It is important to build your vocabulary, especially the words and phrases that you use in flight operations. Remember the English words for things that you deal with during flight. You might not always know the correct word or expression, so be ready to choose a similar word.

- *Practice describing situations!*

  You must be able to describe situations that might occur during flight in plain English. Don't try to memorize responses to specific questions or situations. Use the exercises in the book to practice describing different situations. Pay attention to all phases of flight operations, and become familiar with both normal and non-normal situations that might occur.

- *Remember the 6 areas of evaluation!*

  It is important for you to know what your level of proficiency is. If you have taken the test before, look at the results and work on the areas that were below Level 4. If you have not taken the test yet, work through the book carefully and practice all the exercises. There is a Self-Evaluation checklist on page 106 to help you check your proficiency in each of the 6 areas of evaluation.

- *Use your English!*

  It takes time and hard work to improve your English ability. Don't wait to study! It is important to practice speaking and listening to English every day. Aim for at least 15-30 minutes a day. Make your study time focused and intense. When practicing the exercises in this book, always speak out loud so that you can become used to hearing your voice in English.

INTRODUCTION

本書について

内容

本書には12ユニットが含まれており、各ユニットは、試験のインタビューセクションの3つのステージを基本に構成されています。各ユニットの長さは6ページです。

- 「Introduction」: ユニットで学習するキーワードの紹介；
- 「Single Picture」: インタビューセクションの、「Single Picture」ステージの練習；
- 「ATC Description」: インタビューセクションの、「ATC Description」ステージの練習；
- 「Useful Language」: 役立つ文法構造の紹介；
- 「Picture Sequence」: インタビューセクションの、「Picture Sequence」ステージの練習；
- 「Aviation Reading」: リーディングを通して他の役立つ単語の紹介；

全ユニットを通して、6項目の評価項目に対する練習ができるようになっています。また、インタビューセクションの、異なるステージで役立つアドバイスも用意しました。1～3のユニット、そして7～9のユニットでは、特に日本人が苦手とする、B/V、S/TH、L/R の発音練習を追加してあります。4～6のユニットと、10～12のユニットには、リスニング問題を追加しました。

本書の使い方

- 語彙を増やしましょう！ 語彙を増やすことは重要です。特に航空従事中に使用する単語やフレーズは重要です。フライトに関わる英単語を思い出すようにしてください。正しい単語や表現が分からない場合もあるので、同じような意味を持つ単語を選べるようにしておきます。
- 状況を描写する練習をしましょう！ フライト中の状況を、日常英語で描写できなければなりません。問題や状況を表す回答を暗記することは避けます。本書に登場する異なる状況を描写する練習をします。航空従事の全ての「flight phase」に気を配り、起こり得る通常および非常どちらの状況に対しても親しんでおきましょう。
- 6項目の評価項目を思い出しましょう！ 自身の能力のレベルを知ることは重要です。試験を受験したことがあるのならば、結果を見てレベル4以下だった 項目に対して勉強します。試験を受験したことがない場合は、本書を注意深く熟読して、全てのエクササイズを行い練習しましょう。106 ページにはセルフエバリュエーションがありますから、6つの評価項目「6 Areas of Evaluation」での自身の能力をチェックすることができます。
- 英語を使いましょう！ 英語能力を上げるには、時間と努力が必要です。勉強を待ってはなりません！毎日のスピーキングとリスニング 練習が大切になります。一日に15～30分を目標にしてみてください。勉強時間には集中して中身を濃くするようにしてください。本書のエクササイズを行っている時には、常に声に出して行うようにし、自身の英語の声を聴くことに慣れるようにしてください。

CD-ROMについて

- ファイルは138個あります。本文中の[CD1] [CD2] [CD3]……の表記は、CD-ROM中のファイル名[Unit1 CD001] [Unit1 CD002] [Unit1 CD003]……に対応しています。

Are you ready for departure?

UNIT 1　　　　　　　　　　　　PRE-FLIGHT OPERATIONS
"LADIES AND GENTLEMEN..."
DELAYS DUE TO WEATHER

VOCABULARY
以下のキーワードを読みましょう。分からないキーワードに○をして、意味を調べましょう。

KEY WORDS
1. adverse weather
2. at your discretion
3. delay
4. departure
5. gate
6. hazard
7. holding pattern
8. lightning
9. microburst
10. runway
11. sleet
12. thunderstorm
13. turbulence

ABBREVIATION
1. TWR

PRONUNCIATION
CD1 CDで上のキーワードを聞き、後についてリピートしましょう。何回か繰り返します。

PRONUNCIATION
CD2 CDで以下の言葉を聞き、後についてリピートしましょう。

B/V IN INITIAL POSITION
1. **b**an – **v**an
2. **b**at – **v**at
3. **b**eer – **v**eer
4. **b**ent – **v**ent
5. **b**erry – **v**ery
6. **b**est – **v**est
7. **b**et – **v**et
8. **b**oat – **v**ote
9. **b**olt – **v**olt
10. **b**owl – **v**ole

For "**B**", put your lips together, then open them quickly!

PRONUNCIATION
CD3 CDで以下の表現を聞き、後についてリピートしましょう。

B/V IN INITIAL POSITION
1. **b**ad weather
2. poor **v**isibility
3. **b**oarding **b**ridge
4. **v**ery dangerous
5. **b**ig thunderstorm
6. **v**isual approach
7. **v**icinity of the airport
8. hazard **b**eacon
9. **v**olcanic ash
10. **b**raking action
11. radar **v**ector
12. **v**ariable winds
13. **b**lowing snow

For "**V**", put your top front teeth and bottom lip together, then open your mouth!

FLUENCY
CD4 CDで以下の文を聞き、音読しましょう。何回か音読を繰り返します。

DESCRIBING A SINGLE PICTURE

The weather is very bad at this airport. There is a big thunderstorm, and I can see clouds, strong winds and lightning. This is a very dangerous situation because thunderstorms might create severe turbulence and microbursts. The wind is blowing in different directions, so this may be a microburst. I can see an airplane on the runway. I don't know if the pilot is trying to land or take off, but this situation is too dangerous. Therefore, I think it is safer to wait until the weather gets better.

UNIT 1　　　　　　　　　　　　　　　　　　　PRE-FLIGHT OPERATIONS

SINGLE PICTURE

ADVICE　主題と話題を見つけよう

「SINGLE PICTURE」の問題では、まず主題を見つけるようにしましょう。そうすると、聞き手に話が伝わりやすくなります。それから、それぞれの話題に対する詳細を考えます。絵を見て気がついた点をランダムに挙げるのは避けましょう。回答は、構成されていて聞き手にとって分かりやすくなければなりません。

X　"I see clouds."　　**O**　"The weather is very bad. There is a thunderstorm approaching the airport."

FLUENCY

絵の状況を描写しましょう。まず、絵の中から3つか4つの話題を見つけます（例えば、weather、holding、delays等）。次にそれぞれの話題に関して1分程度話します。目標は、3分間話すこと、完全文を使うこと、長い沈黙を避けること、er、you know、wellなどのつなぎ語を避けることです。

INTERACTIONS

CD5 上の絵について以下の問題に答えましょう。まずは、CDでそれぞれの問題を聞きます。問題を聞いた後に一時停止ボタンを押して、それぞれの問題に対して約1分間で回答します。回答は声に出して言いましょう。

FOLLOW-UP QUESTIONS

1. How may the adverse weather in this picture affect departures and arrivals?
2. What might happen if an aircraft is struck by lightning?
3. What happens if the electrical power goes out at the airport?

UNIT 1   PRE-FLIGHT OPERATIONS

PICTURE SEQUENCE

Identify the topics, then add details!   *Try to make simple sentences!*

FLUENCY

1～4の順に、絵の状況を描写しましょう。
目標は、3分間話すこと、完全文を使うこと、
長い沈黙を避けること、er、you know、wellなどのつなぎ語を避けることです。

INTERACTIONS

CD10 上の絵について以下の問題に答えましょう。まずは、CDでそれぞれの問題を聞きます。問題を聞いた後に一時停止ボタンを押して、それぞれの問題に対して約1分間で回答します。回答は声に出して言いましょう。

FOLLOW-UP QUESTIONS

1. Why were the airplanes in a holding pattern?
2. What happens to the passengers if a flight is cancelled?
3. How long can the flight crew stand by and wait for the departure?

UNIT 1　　　　　　　　　　　　　　　　　PRE-FLIGHT OPERATIONS

AVIATION READING

VOCABULARY

以下のキーワードを読みましょう。分からないキーワードに〇をして、意味を調べましょう。

KEY WORDS

1. accident
2. commercial airliner
3. crosswind
4. electrical surge
5. en route
6. fuel tank
7. fuel vapor
8. ignite
9. lightning strike

ACRONYM

1. FAA

PRONUNCIATION

CD11 CDで上のキーワードを聞き、後についてリピートしましょう。何回か繰り返します。

FLUENCY

CD12 CDで以下の文を聞き、音読しましょう。何回か音読を繰り返します。

LIGHTNING STRIKE

　The average commercial airliner is struck by lightning approximately once a year. Fortunately, there have been very few accidents caused by lightning in recent decades, but in the early period of commercial jet aviation it was a serious hazard.

　On December 8th 1963, a Pan American Boeing 707 was en route from Baltimore to Philadelphia in the United States when it crashed after being hit by lightning. The aircraft was in a holding pattern when lightning struck the left wing, igniting fuel vapor in one of the fuel tanks. This caused an explosion which destroyed the outer part of the wing. All 81 passengers and crew were killed in the subsequent crash. As a result of this accident, the Federal Aviation Administration (FAA) ordered that all commercial jets in the United States be fitted with lightning discharge equipment.

　Modern airliners are designed to survive lightning strikes. Electronics and navigation equipment are protected from electrical surges, and fuel tanks are tested to ensure that they can withstand a lightning strike. However, this does not mean that pilots can fly safely through thunderstorms. Other weather hazards associated with thunderstorms – such as severe turbulence, icing conditions and strong crosswinds – may be dangerous for even the most modern aircraft.

UNIT 2 — AT THE RAMP
"USE CAUTION FOR ICY CONDITIONS"
SNOW & ICE AT THE AIRPORT

VOCABULARY
以下のキーワードを読みましょう。分からないキーワードに○をして、意味を調べましょう。

KEY WORDS

1. accumulate
2. anti-icing
3. boarding stairs
4. cancellation
5. cargo
6. clearance
7. de-icing
8. fuselage
9. precaution
10. pre-flight inspection
11. ramp
12. snow removal
13. spot
14. taxiway
15. winter operations

PRONUNCIATION
CD13 CDで上のキーワードを聞き、後についてリピートしましょう。何回か繰り返します。

PRONUNCIATION
CD14 CDで以下の言葉を聞き、後についてリピートしましょう。

S/TH IN INITIAL POSITION

1. **s**ank – **th**ank
2. **s**aw – **th**aw
3. **s**ick – **th**ick
4. **s**igh – **th**igh
5. **s**in – **th**in
6. **s**ing – **th**ing
7. **s**ink – **th**ink
8. **s**ome – **th**umb
9. **s**ong – **th**ong
10. **s**ort – **th**ought

For "**S**", put the tip of your tongue behind the bottom front teeth!

PRONUNCIATION
CD15 CDで以下の表現を聞き、後についてリピートしましょう。

S/TH IN INITIAL POSITION

1. heavy **s**nowstorm
2. reverse **th**rust
3. taxiway **s**ign
4. **s**trong crosswind
5. big **th**understorm
6. **s**lippery taxiway
7. **th**ree **s**ix right
8. **s**tall **s**peed
9. displaced **th**reshold
10. **s**evere turbulence
11. **th**ermal cell
12. full **th**rottle
13. dangerous **s**ituation

For "**TH**", put the tip of your tongue between the top and bottom front teeth!

FLUENCY
CD16 CDで以下の文を聞き、音読しましょう。何回か音読を繰り返します。

DESCRIBING A SINGLE PICTURE

There is adverse weather at this airport. It looks like winter operations because there is snow everywhere. I can see several vehicles on the ramp. Snow removal equipment is working to clear the snow. The airplane has passenger boarding stairs attached, but I think it is not ready for departure because too much snow and ice has accumulated on the wings and fuselage. This is a dangerous situation. The airplane must be de-iced before it can take off.

UNIT 2　　　　　　　　　　　　　　　　　　　　AT THE RAMP

SINGLE PICTURE

Identify the topics, then add details!

Remember to make simple sentences!

FLUENCY

絵の状況を描写しましょう。まず、絵の中から3つか4つの話題を見つけます（例えば、de-icing、delays、snow等）。次にそれぞれの話題に関して1分程度話します。目標は、3分間話すこと、完全文を使うこと、長い沈黙を避けること、er、you know、wellなどのつなぎ語を避けることです。

INTERACTIONS

CD17 上の絵について以下の問題に答えましょう。まずは、CDでそれぞれの問題を聞きます。問題を聞いた後に一時停止ボタンを押して、それぞれの問題に対して約1分間で回答します。回答は声に出して言いましょう。

FOLLOW-UP QUESTIONS

1. What is being done to make this airport safe for flight operations?
2. What special preparations must pilots make for these weather conditions?
3. How does ice and snow on the wings and fuselage affect aircraft performance?

UNIT 2 AT THE RAMP

ATC DESCRIPTION

INTERACTIONS

CD18 下の絵について以下の問題に答えましょう。まずは、CDでそれぞれの問題を聞きます。問題を聞いた後に一時停止ボタンを押して、それぞれの問題に対して約1分間で回答します。回答は声に出して言いましょう。

WARM-UP QUESTIONS

1. Describe the situation in the picture.
2. Is this airplane ready for departure? Why (not)?
3. What actions should the pilots take?
4. What help do the pilots need from the ground staff?

COMPREHENSION

CD19 CDでATCの会話を聞き、正しい答えを選びましょう。(回答は91ページ)

ATC DIALOG

1. The delay is due to…
   (a) a tow truck   (b) ice on the taxiway   (c) snow on the runway   (d) a vehicle on the taxiway
2. The estimate for the delay is…
   (a) unknown   (b) several minutes   (c) about 15 minutes   (d) about 30 minutes
3. The pilot decides to…
   (a) cancel the flight   (b) return to the ramp   (c) hold his position   (d) request takeoff clearance

FLUENCY

上のATCの状況を描写しましょう。1～2分話すようにします。

INTERACTIONS

CD20 上のATCの状況について、以下の問題に答えましょう。まずは、CDでそれぞれの問題を聞きます。問題を聞いた後に一時停止ボタンを押して、それぞれの問題に対して約1分間で回答します。回答は声に出して言いましょう。

FOLLOW-UP QUESTIONS

1. Describe some common types of delays at the gate and ramp area.
2. Name some situations that would cause a flight cancellation.
3. What does 'no show' passenger mean? What would you do in that situation?

UNIT 2   AT THE RAMP

USEFUL LANGUAGE

STRUCTURE

このエクササイズでは、単純過去形を使って、過去に起きた事を表わします。下の単語を並べ替えて文を作りましょう。1問目の例を参考にしてください。

EVENTS IN THE PAST

1. takeoff. / before / de-iced / the aircraft / They → _They de-iced the aircraft before takeoff._
2. slowly / the ice. / of / The pilot / because / taxied → _____
3. decided / The captain / to / the flight. / cancel → _____
4. for / continued / a long time. / to / Snow / fall → _____
5. removed / the runway. / from / snow / Workers → _____
6. takeoff. / the cabin / Cabin crew / before / secured → _____
7. the plane. / Airport staff / onto / cargo / loaded → _____
8. system. / activate / to / They / the anti-icing / decided → _____

PRONUNCIATION   CD21 CDで上の文を聞き、後についてリピートしましょう。何回か繰り返します。(回答は91ページ)

STRUCTURE

下の絵は、エプロンでのさまざまな作業を描写しています。下の「USEFUL STRUCTURE」とそれぞれの絵の上の単語を使って、起きた事を表わす文を作りましょう。1問目の例を参考にしてください。

USEFUL STRUCTURE

SUBJECT	SIMPLE PAST TENSE
The captain	decided…
Airport staff	removed…
They	checked…
etc	etc

2 walk-around • inspect • tires

3 cargo door • load • container

1 open spot • accumulate • remove

Use past tenses for the picture sequence!

4 dog • cargo • dangerous goods

Last week, a jet aircraft <u>was</u> at an open spot.
Snow and ice <u>accumulated</u> on the wings and fuselage.
Airport staff <u>removed</u> the snow from the ramp.

UNIT 2   AT THE RAMP

PICTURE SEQUENCE

ADVICE 過去形を保って ／ 絵のつながりに気を付けよう

「PICTURE SEQUENCE」の問題では、過去に起きた状況を伝えるために、つねに過去形で話しましょう。また、それぞれの絵のつながりを明確にして、話が伝わりやすくなるように気を付けます（例：and、but、so、becauseを使う）。

X "There is snow. I see an airplane."   **O** "Last week, there was a heavy snowfall, so airport workers removed the snow."

FLUENCY   1〜4の順に、絵の状況を描写しましょう。目標は、3分間話すこと、完全文を使うこと、長い沈黙を避けること、er、you know、wellなどのつなぎ語を避けることです。

INTERACTIONS   **CD22** 上の絵について以下の問題に答えましょう。まずは、CDでそれぞれの問題を聞きます。問題を聞いた後に一時停止ボタンを押して、それぞれの問題に対して約1分間で回答します。回答は声に出して言いましょう。

FOLLOW-UP QUESTIONS

1. How does a pilot get information about weather?
2. What is the difference between de-icing and anti-icing?
3. What precautions do you use when flying in adverse weather conditions?

UNIT 2 AT THE RAMP

AVIATION READING

VOCABULARY

以下のキーワードを読みましょう。分からないキーワードに○をして、意味を調べましょう。

KEY WORDS
1. contamination
2. factor
3. gain altitude
4. heavy snowstorm
5. marginal conditions
6. reverse thrust
7. snow chains
8. takeoff roll
9. tow tractor

ACRONYM
1. NTSB

PRONUNCIATION CD23
CDで上のキーワードを聞き、後についてリピートしましょう。何回か繰り返します。

FLUENCY CD24
CDで以下の文を聞き、音読しましょう。何回か音読を繰り返します。

WE'RE GOING DOWN!

Washington National Airport in the United States was closed by a heavy snowstorm on the morning of January 13th 1982. It reopened at midday under marginal conditions, but snow continued to fall. Air Florida Flight 90, a Boeing 737, had difficulty leaving the gate because the tow tractor could not get traction on the ice. The crew tried unsuccessfully to move away using engine reverse thrust. Finally, a tow tractor equipped with snow chains pushed the plane back from the gate.

Many aircraft were waiting in the taxi line, and as a result it took almost 50 minutes for Flight 90 to reach the takeoff runway. During this time, more snow and ice accumulated on the wings. The captain realised this, but decided not to return to the gate for more de-icing because he did not want a further delay. Furthermore, the crew decided not to activate the engine anti-icing system, which meant that engine sensors were providing inaccurate readings.

Heavy snow was falling during the takeoff roll, and the aircraft struggled to gain altitude because of snow and ice contamination on the wings. Less than one minute later it struck the 14th Street Bridge and crashed into the Potomac River, killing 74 of the passengers and crew as well as four people on the bridge. Four passengers and one cabin attendant were rescued from the icy river.

The National Transportation Safety Board (NTSB) investigation found that many factors contributed to this accident. These factors included the crew's failure to use engine anti-icing and their decision to take off with snow and ice on the wings.

UNIT 3 — GROUND MOVEMENT

"HOLD YOUR POSITION!"
OBSTRUCTIONS ON THE TAXIWAY

VOCABULARY

以下のキーワードを読みましょう。分からないキーワードに○をして、意味を調べましょう。

KEY WORDS
1. baggage
2. baggage car
3. braking action
4. briefing
5. hold your position
6. hydraulic fluid
7. leak
8. obstruction
9. run-up area
10. slippery
11. taxi

ABBREVIATION
1. RWY

ACRONYM
1. PAPI

PRONUNCIATION

CD25 CDで上のキーワードを聞き、後についてリピートしましょう。何回か繰り返します。

PRONUNCIATION

CD26 CDで以下の言葉を聞き、後についてリピートしましょう。

L/R IN INITIAL POSITION
1. **l**ack – **r**ack
2. **l**ake – **r**ake
3. **l**amp – **r**amp
4. **l**and – **r**and
5. **l**ane – **r**ain
6. **l**ate – **r**ate
7. **l**ead – **r**ead
8. **l**eak – **r**eek
9. **l**ift – **r**ift
10. **l**ight – **r**ight
11. **l**oad – **r**oad
12. **l**ong – **wr**ong
13. **l**ow – **r**ow

For "*L*", touch the tip of your tongue to the top of your mouth!

PRONUNCIATION

CD27 CDで以下の表現を聞き、後についてリピートしましょう。

L/R IN INITIAL POSITION
1. flight **l**evel
2. pilot **r**esponsibility
3. **l**anding gear
4. **r**amp area
5. **l**eaking engine
6. **r**ate of turn
7. **l**oad factor
8. **r**outine inspection
9. **r**udder pedals
10. PAPI **l**ights
11. **r**unway intrusion
12. **r**ight of way
13. weather **l**imitations

For "*R*", lift the middle of your tongue but don't touch the top of your mouth!

FLUENCY

CD28 CDで以下の文を聞き、音読しましょう。何回か音読を繰り返します。

DESCRIBING A SINGLE PICTURE

There seems to be some hydraulic fluid on the taxiway. Hydraulic fluid is a kind of obstruction. Sometimes it leaks from the airplane. It can be dangerous because it makes the taxiway slippery and the braking action may become poor or nil. The fluid must be removed. If I see fluid on the runway or taxiway, I will first hold my position and then report it to the controller.

UNIT 3 GROUND MOVEMENT

SINGLE PICTURE

ADVICE 現在形を保って ／ 完全文で話そう

「SINGLE PICTURE」の問題では、今見えている状況や起きている事を描写しますから、つねに現在形で話しましょう。また、完全文で話すようにします。一言で回答したり不完全文を使うと、意味が伝わりにくくなったり、理解されにくくなってしまいます。

X "Airplane taxi."　　O "The pilot is taxiing on the taxiway."

FLUENCY　絵の状況を描写しましょう。まず、絵の中から3つか4つの話題を見つけます（例えば、taxi、birds、obstruction等）。次にそれぞれの話題に関して1分程度話します。目標は、3分間話すこと、完全文を使うこと、長い沈黙を避けること、er、you know、wellなどのつなぎ語を避けることです。

INTERACTIONS　**CD29** 上の絵について以下の問題に答えましょう。まずは、CDでそれぞれの問題を聞きます。問題を聞いた後に一時停止ボタンを押して、それぞれの問題に対して約1分間で回答します。回答は声に出して言いましょう。

FOLLOW-UP QUESTIONS

1. Name some obstructions that may be seen on a taxiway.
2. What action should pilots take if there is an obstruction during taxi?
3. What should pilots do if their aircraft hits an obstruction during taxi?

UNIT 3   GROUND MOVEMENT

ATC DESCRIPTION

INTERACTIONS

CD30 下の絵について以下の問題に答えましょう。まずは、CDでそれぞれの問題を聞きます。問題を聞いた後に一時停止ボタンを押して、それぞれの問題に対して約1分間で回答します。回答は声に出して言いましょう。

WARM-UP QUESTIONS

1. Describe the situation in the picture.
2. What is the pilot doing? Why?
3. How do you think the obstruction got on the taxiway?
4. What will be done to remove the obstruction?

COMPREHENSION

CD31 CDでATCの会話を聞き、下線部を埋めましょう。(回答は92ページ)

ATC DIALOG

PILOT: Tokyo Ground, OB Air 1663, we are __1._____ at Taxiway Oscar. There seems to be some kind of obstruction in front of us.

CONTROLLER: OB Air 1663, Tokyo Ground, can you describe the __2._____ for us?

PILOT: OB Air 1663, it looks like some kind of a box, a large rectangular box that might have fallen off a __3._____.

CONTROLLER: OB Air 1663, can you move __4._____?

PILOT: Negative, OB Air 1663.

CONTROLLER: Ok, OB Air 1663, hold your __5._____ until further notice. We are sending a truck to pick it up. Expect a delay of about __6._____.

FLUENCY

上のATCの状況を描写しましょう。1～2分話すようにします。

INTERACTIONS

CD32 上のATCの状況について、以下の問題に答えましょう。まずは、CDでそれぞれの問題を聞きます。問題を聞いた後に一時停止ボタンを押して、それぞれの問題に対して約1分間で回答します。回答は声に出して言いましょう。

FOLLOW-UP QUESTIONS

1. How would you handle this situation?
2. How could this situation have been avoided?
3. Have you ever had a similar experience? If so, describe what happened.

UNIT 3   GROUND MOVEMENT

USEFUL LANGUAGE

STRUCTURE

このエクササイズでは、「some kind of 」「seems to be」「looks like」の表現を使って、障害物を表わします。下の単語を並べ替えて文を作りましょう。1問目の例を参考にしてください。

DESCRIBING OBSTRUCTIONS

1. seems / baggage. / It / a piece of / to be → _It seems to be a piece of baggage._
2. It / the taxiway. / looks / on / hydraulic fluid / like → _____
3. RWY 34L. / to be / seems / on / There / an animal → _____
4. looks / It / plastic. / it is / like / made of → _____
5. tall. / about / seems / It / 1 metre / to be → _____
6. like / ahead. / there is / looks / It / a big piece of metal → _____
7. There / to be / in front of us. / an obstruction / seems → _____
8. container / is / ahead of us. / There / some kind of → _____

PRONUNCIATION

CD33 CDで上の文を聞き、後についてリピートしましょう。何回か繰り返します。（回答は92ページ）

STRUCTURE

下の絵は、誘導路のさまざまな障害物を描写しています。下の「USEFUL STRUCTURE」とそれぞれの絵の上の単語を使って、障害物を表わす文を作りましょう。1問目の例を参考にしてください。

USEFUL STRUCTURE

There is some kind of…
There seems to be…
It seems to be…
It looks like…

2 barrel • 1m tall • metal

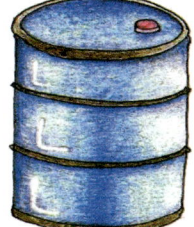

5 bag • 120cm tall • golf clubs

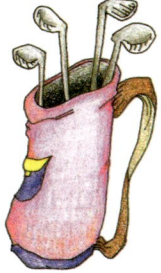

1 tool • 20cm long • metal

3 suitcase • 70cm wide • leather

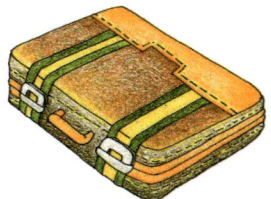

6 trash can • 1m tall • metal

There is some kind of tool on the taxiway.

It seems to be about 20 centimeters long.

It looks like it is made of metal, and the color is silver.

4 hammer • 30cm long • wood & metal

Try to make complete sentences!

UNIT 3   GROUND MOVEMENT

PICTURE SEQUENCE

Use past tenses for the picture sequence!

Make connections between the pictures!

FLUENCY

1〜4の順に、絵の状況を描写しましょう。
目標は、3分間話すこと、完全文を使うこと、
長い沈黙を避けること、er、you know、wellなどのつなぎ語を避けることです。

INTERACTIONS

CD34 上の絵について以下の問題に答えましょう。まずは、CDでそれぞれの問題を聞きます。問題を聞いた後に一時停止ボタンを押して、それぞれの問題に対して約1分間で回答します。回答は声に出して言いましょう。

FOLLOW-UP QUESTIONS

1. Was the plane arriving at the airport? How do you know this?
2. Describe exactly how the pilot gets from the ramp to the run-up area.
3. What are some other possible reasons for flight delays? Describe your experience.

UNIT 3   GROUND MOVEMENT

AVIATION READING

VOCABULARY
以下のキーワードを読みましょう。分からないキーワードに〇をして、意味を調べましょう。

KEY WORDS
1. accelerate
2. bank
3. burst resistant
4. damage
5. debris
6. fire warning
7. port wing
8. shut down

ACRONYMS
1. FOD
2. V_1

PRONUNCIATION
CD35 CDで上のキーワードを聞き、後についてリピートしましょう。何回か繰り返します。

FLUENCY
CD36 CDで以下の文を聞き、音読しましょう。何回か音読を繰り返します。

DEBRIS ON THE RUNWAY

On July 25th 2000, a Continental Airlines DC-10 lost a piece of metal during takeoff from Charles de Gaulle Airport in Paris, France. This piece of foreign object debris (FOD), about 3cm wide by 43cm long, was still on the runway when an Air France Concorde began its takeoff roll.

One of the Concorde's tires hit the debris and burst, causing damage to the aircraft's wing fuel tanks. Leaking fuel ignited, and flames could be seen coming from the aircraft's port wing.

As the Concorde had passed V_1, the crew continued with the takeoff. However, engine two had been shut down in response to a fire warning, and the aircraft could not climb or accelerate.

Meanwhile, fire damage to the port wing and the failure of engine one resulted in the aircraft banking more than 100 degrees. The crew reduced power on engines three and four but were unable to regain control of the aircraft, which crashed into a hotel near the airport.

All 100 passengers and 9 crew died in the crash, as well as 4 people on the ground. Following this accident, changes were made to Concorde, including burst resistant tires. The aircraft type never fully recovered, though, and commercial Concorde flights stopped in 2003.

UNIT 4　　　CLEARED FOR TAKEOFF

"STAND BY FOR CLEARANCE"
REJECTING TAKEOFF

VOCABULARY
以下のキーワードを読みましょう。分からないキーワードに○をして、意味を調べましょう。

KEY WORDS
1. bird strike
2. bird sweep
3. evacuate
4. fire equipment
5. go around
6. hold
7. incident
8. landing gear
9. puncture
10. rejected takeoff
11. re-schedule
12. stand by
13. tow assistance

ABBREVIATION
1. T/O

ACRONYMS
1. FBO
2. PF

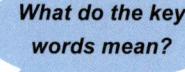
What do the key words mean?
Try to explain in English!

PRONUNCIATION
CD37 CDで上のキーワードを聞き、後についてリピートしましょう。何回か繰り返します。

COMPREHENSION
CD38 CDで5つのATCの会話を聞き、それぞれの問題に答えましょう。（回答は93ページ）

LISTENING TO ATC
1. (a) Why is RWY 34 closed? (b) How long do the pilots have to wait?
2. (a) What does the tower tell King Air 12F to do? (b) What is the reason?
3. (a) What does King Air 12F want to do? (b) Why doesn't King Air 12F want to take off now?
4. (a) What instructions has King Air 12F received? (b) What is King Air 12F doing now?
5. (a) Which runway can King Air 12F take off from? (b) Why does the tower want King Air 12F to go straight ahead after takeoff?

FLUENCY
CD39 CDで以下の文を聞き、音読しましょう。何回か音読を繰り返します。

DESCRIBING A SINGLE PICTURE

This is the landing gear of a large jet aircraft. It looks like the landing gear tires are flat or punctured. Maybe the aircraft ran over an obstruction at the ramp or on the taxiway. This aircraft is not ready for take-off. It cannot move on its own power so the pilot must request tow assistance from the company dispatch or the airport FBO. The repair will take time so I think that the flight has to be delayed or canceled. The passengers must get off the airplane and be re-scheduled to another flight.

UNIT 4   CLEARED FOR TAKEOFF

SINGLE PICTURE

Use present tenses for the single picture!

Remember to use complete sentences!

FLUENCY

絵の状況を描写しましょう。まず、絵の中から3つか4つの話題を見つけます（例えば、departure、birds、takeoff等）。次にそれぞれの話題に関して1分程度話します。目標は、3分間話すこと、完全文を使うこと、長い沈黙を避けること、er、you know、wellなどのつなぎ語を避けることです。

INTERACTIONS

CD40 上の絵について以下の問題に答えましょう。まずは、CDでそれぞれの問題を聞きます。問題を聞いた後に一時停止ボタンを押して、それぞれの問題に対して約1分間で回答します。回答は声に出して言いましょう。

FOLLOW-UP QUESTIONS

1. What precautions should the pilot entering the runway take?
2. Why are birds dangerous to pilots?
3. How are birds removed from the airport area?

UNIT 4 CLEARED FOR TAKEOFF

ATC DESCRIPTION

INTERACTIONS

CD41 下の絵について以下の問題に答えましょう。まずは、CDでそれぞれの問題を聞きます。問題を聞いた後に一時停止ボタンを押して、それぞれの問題に対して約1分間で回答します。回答は声に出して言いましょう。

WARM-UP QUESTIONS

1. Describe the situation in the picture.
2. Name some reasons for a rejected takeoff.
3. What might happen if a pilot tries to reject takeoff after V_1?
4. What actions should the pilot take after rejecting takeoff?

COMPREHENSION

CD42 CDでATCの会話を聞きながら3つの間違いに○をして、正しい言葉を書きましょう。
（回答は93ページ）

ATC DIALOG

CONTROLLER:	OB Air 1663, Tokyo TWR, use caution, birds in the vicinity of the airport, wind 030 at 10, RWY 34L, cleared for T/O.
PILOT 1:	Tokyo TWR, OB Air 1663, cleared for T/O, RWY 34L, starting roll.
PILOT 1:	Tokyo TWR, OB Air 1663, rejecting T/O, it seems we have hit an object on the runway, request hold on runway.
CONTROLLER:	OB Air 1663, roger, understand your situation, hold your position, RWY 34L.
PILOT 1:	OB Air 1663, holding on RWY 34L.
CONTROLLER:	Park Air 7430, Tokyo TWR, go around RWY 34L, aircraft on the runway.
PILOT 2:	Roger, Park Air 7430 is going around.
PILOT 1:	Tokyo TWR, OB Air 1663, we are going to need a tow. It seems we have a locked brake.
CONTROLLER:	Roger, OB Air 1663, standby for your request.

FLUENCY

上のATCの状況を描写しましょう。1〜2分話すようにします。

INTERACTIONS

CD43 上のATCの状況について、以下の問題に答えましょう。まずは、CDでそれぞれの問題を聞きます。問題を聞いた後に一時停止ボタンを押して、それぞれの問題に対して約1分間で回答します。回答は声に出して言いましょう。

FOLLOW-UP QUESTIONS

1. Why did OB Air 1663 reject the takeoff?
2. What must be done before the runway can become active again?
3. Have you ever experienced a rejected takeoff? If so, describe what happened.

UNIT 4   CLEARED FOR TAKEOFF

USEFUL LANGUAGE

STRUCTURE

このエクササイズでは、助動詞の「must」と「has/have to」を使って、必要な行動を表わします。下の単語を並べ替えて文を作りましょう。1問目の例を参考にしてください。

NECESSARY ACTIONS

1. have / their position. / The pilots / hold / to → *The pilots have to hold their position.*
2. remove / Airport / the birds. / must / staff → _____
3. tow / has / The captain / to / assistance. / request → _____
4. canceled. / must / flight / be / The → _____
5. They / wait / is active. / the runway / have to / until → _____
6. ahead. / continue / must / PF / straight / The → _____
7. get / have / the airplane. / to / The passengers / off → _____
8. 30 minutes. / be closed / to / for / The runway / has → _____

PRONUNCIATION

CD44 CDで上の文を聞き、後についてリピートしましょう。何回か繰り返します。(回答は93ページ)

STRUCTURE

下の絵は、離陸時のさまざまな問題を描写しています。下の「USEFUL STRUCTURE」とそれぞれの絵の上の単語を使って、必要な対応を表わしましょう。1問目の例を参考にしてください。

USEFUL STRUCTURE

SUBJECT	MODAL	VERB
The captain	has to	notify…
The pilots	have to	request…
They	must	wait…
etc		etc

2 tire • puncture • tow

3 leak • hydraulic fluid • clean up

1 tower • hold • go around

Use simple, accurate sentences!

4 engine • fire equipment • evacuate

The pilots <u>must</u> notify the tower about the rejected takeoff.
They <u>have to</u> request a hold on the runway.
The tower <u>has to</u> instruct other traffic to go around.

31

UNIT 4   CLEARED FOR TAKEOFF

PICTURE SEQUENCE

ADVICE 時制をコントロールしよう

「PICTURE SEQUENCE」の問題では、時制に気を付けることが大変重要です。過去形、現在形、未来形を行ったり来たりするのは避けましょう。過去に起きた事柄を話している事を忘れずに。

X "One day last week, I see birds on the runway."   O "One day last week, I saw birds on the runway."

FLUENCY

1〜4の順に、絵の状況を描写しましょう。目標は、3分間話すこと、完全文を使うこと、長い沈黙を避けること、er、you know、wellなどのつなぎ語を避けることです。

INTERACTIONS

CD45 上の絵について以下の問題に答えましょう。まずは、CDでそれぞれの問題を聞きます。問題を聞いた後に一時停止ボタンを押して、それぞれの問題に対して約1分間で回答します。回答は声に出して言いましょう。

FOLLOW-UP QUESTIONS

1. How could this incident have been prevented?
2. What might have happened if the pilot continued taking off after the bird strike?
3. What are some bird-control measures at your airport?

UNIT 4

CLEARED FOR TAKEOFF

AVIATION READING

VOCABULARY

以下のキーワードを読みましょう。分からないキーワードに○をして、意味を調べましょう。

KEY WORDS
1. abort
2. deceleration
3. decision speed
4. fatalities
5. freighter
6. inboard
7. inquiry
8. knot
9. loss of power
10. overrun
11. situational awareness
12. speed brake
13. starboard

ACRONYM
1. CVR

PRONUNCIATION
CD46 CDで上のキーワードを聞き、後についてリピートしましょう。何回か繰り返します。

FLUENCY
CD47 CDで以下の文を聞き、音読しましょう。何回か音読を繰り返します。

BIRD STRIKE

Bird strikes are a common problem for pilots. In fact, many commercial airline pilots who fly frequently to airports located near water have experienced a bird strike. Bird strikes generally cause little damage to large aircraft, but in some cases they can cause serious damage and even fatalities.

On May 25th 2008, a Kalitta Air Boeing 747 freighter suffered a bird strike during a takeoff roll at Brussels National Airport in Belgium. The crew aborted takeoff, even though the aircraft had already passed V_1. With neither speed brakes nor reverse thrust deployed, the aircraft overran the runway by 300 metres and broke into three pieces. There were no injuries to the four crew members and one passenger on board.

Investigators discovered that a bird had struck the starboard inboard engine as the aircraft accelerated along the runway. The cockpit voice recorder (CVR) revealed that four seconds after the jet reached the V_1 decision speed, there was a loud bang followed by a loss of power from the engine.

The inquiry found that the accident was caused by the decision to reject takeoff 12 knots after passing V_1. In addition, a number of other factors contributed to the accident. These included the situational awareness of the crew and their failure to use some deceleration devices. The inquiry recommended modifications to Kalitta Air training for rejected takeoffs, as well as changes in bird control methods at the airport.

UNIT 5

TAKEOFF & CLIMB

"CHECK AND CONFIRM"
AIR TURNBACK

VOCABULARY

以下のキーワードを読みましょう。分からないキーワードに〇をして、意味を調べましょう。

KEY WORDS

1. air turnback
2. baggage compartment
3. caution
4. equipment failure
5. flap
6. injured passenger
7. nose gear
8. panel
9. retract position
10. sick passenger
11. stuck landing gear

ACRONYM

1. EICAS

What do the key words mean?

Try to explain in English!

PRONUNCIATION

CD48 CDで上のキーワードを聞き、後についてリピートしましょう。何回か繰り返します。

COMPREHENSION

CD49 CDで5つのATCの会話を聞き、それぞれの問題に答えましょう。(回答は94ページ)

LISTENING TO ATC

1. (a) What seems to be the problem? (b) What does the pilot want to do?
2. (a) What must the pilot do after takeoff? (b) What does the pilot have to look out for?
3. (a) What does the pilot want to do? (b) Describe the passenger.
4. (a) Why is King Air 12F stopped on the taxiway? (b) What do you think ground control will do?
5. (a) What are the ground instructions? (b) Describe the taxiway condition.

FLUENCY

CD50 CDで以下の文を聞き、音読しましょう。何回か音読を繰り返します。

DESCRIBING A SINGLE PICTURE

It looks like the aircraft is returning to the departure airport. There are many reasons why a pilot may request an air turnback. One reason might be that there is a minor mechanical difficulty, such as a stuck landing gear or some kind of equipment failure. Another common reason for an air turnback is that there is a problem in the cabin. There may be a sick or injured passenger. For example, if overhead baggage compartments are not closed and locked during takeoff and climb, turbulence may cause bags to fall and passengers may be injured.

UNIT 5   TAKEOFF & CLIMB

SINGLE PICTURE

ADVICE 情報を提供しよう

「SINGLE PICTURE」の問題では、十分に情報を提供することに気を付けて、聞き手が状況を理解できるようにしましょう。回答は、明瞭、正確で、遅れすぎないようにします。短くシンプルな文を用いると、意味が明確に伝わりやすくなります。

X "There is a door."   **O** "There is an emergency door that is not locked."

FLUENCY

絵の状況を描写しましょう。まず、絵の中から3つか4つの話題を見つけます（例えば、cockpit、EICAS、weather等）。次にそれぞれの話題に関して1分程度話します。目標は、3分間話すこと、完全文を使うこと、長い沈黙を避けること、er、you know、wellなどのつなぎ語を避けることです。

INTERACTIONS

CD51 上の絵について以下の問題に答えましょう。まずは、CDでそれぞれの問題を聞きます。問題を聞いた後に一時停止ボタンを押して、それぞれの問題に対して約1分間で回答します。回答は声に出して言いましょう。

FOLLOW-UP QUESTIONS

1. What is an EICAS warning?
2. What options do the pilots have in this situation?
3. The weather looks fine. Should the flight continue? Why (not)?

UNIT 5   TAKEOFF & CLIMB

ATC DESCRIPTION

INTERACTIONS

CD52 下の絵について以下の問題に答えましょう。まずは、CDでそれぞれの問題を聞きます。問題を聞いた後に一時停止ボタンを押して、それぞれの問題に対して約1分間で回答します。回答は声に出して言いましょう。

WARM-UP QUESTIONS

1. Describe the situation in the picture.
2. Why could this be a dangerous situation?
3. What actions should the pilot take?
4. What should the pilot tell the passengers?

COMPREHENSION

CD53 CDでATCの会話を聞き、正しい答えを選びましょう。(回答は94ページ)

ATC DIALOG

1. OB Air 1663 is instructed to…
   (a) climb to 8,000   (b) contact Center on 126.6   (c) land on RWY 34L   (d) contact TWR for clearance
2. OB Air 1663 wants to…
   (a) turn to the left   (b) return to the airport   (c) descend to 1,500   (d) declare an emergency
3. The problem seems to be…
   (a) lost position   (b) a stuck landing gear   (c) incorrect clearance   (d) a flap in the wrong position

FLUENCY

上のATCの状況を描写しましょう。1〜2分話すようにします。

INTERACTIONS

CD54 上のATCの状況について、以下の問題に答えましょう。まずは、CDでそれぞれの問題を聞きます。問題を聞いた後に一時停止ボタンを押して、それぞれの問題に対して約1分間で回答します。回答は声に出して言いましょう。

FOLLOW-UP QUESTIONS

1. How would you handle this situation?
2. What conditions would make this situation an emergency?
3. What other mechanical problems might happen after takeoff?

UNIT 5   **TAKEOFF & CLIMB**

USEFUL LANGUAGE

STRUCTURE

このエクササイズでは、現在進行形の「is/are VERB+ing」を使って、今起きている事を表わします。下の単語を並べ替えて文を作りましょう。1問目の例を参考にしてください。

EVENTS HAPPENING NOW

1. A red / light / flashing. / is / warning → _A red warning light is flashing._
2. around / the / flying / are / airport. / Birds → _____
3. engine / experiencing / The airplane / problems. / is → _____
4. is / traffic. / A dead / obstructing / animal → _____
5. the taxiway. / are / The pilots / holding / on → _____
6. down / The captain / the engine. / shutting / is → _____
7. the departure / is / to / airport. / The plane / returning → _____
8. taxiway. / approaching / is / on / A Learjet / the → _____

PRONUNCIATION

CD55 CDで上の文を聞き、後についてリピートしましょう。何回か繰り返します。(回答は94ページ)

STRUCTURE

下の絵は、離陸後のさまざまな問題を描写しています。下の「USEFUL STRUCTURE」とそれぞれの絵の上の単語を使って、今起きている事を表わしましょう。1問目の例を参考にしてください。

USEFUL STRUCTURE

SUBJECT	PRESENT CONTINUOUS TENSE	
The aircraft	is	experiencing…
The pilots	are	notifying…
The controller		requesting…
etc		etc

2 panel • fall off • check

3 climb • nose gear • fix

1 flap • ATC • solve problem

Use present tenses for the single picture!

4 smoke • engine • emergency

This aircraft <u>is experiencing</u> a flap problem.
I think the pilots <u>are notifying</u> ATC.
They <u>are working</u> together to solve the problem.

UNIT 5　　　　　　　　　　　　　　　　　　　　　　　　　TAKEOFF & CLIMB

PICTURE SEQUENCE

Identify the topics, then add details!

Remember to control your tenses!

FLUENCY

1～4の順に、絵の状況を描写しましょう。
目標は、3分間話すこと、完全文を使うこと、
長い沈黙を避けること、er、you know、wellなどのつなぎ語を避けることです。

INTERACTIONS

CD56 上の絵について以下の問題に答えましょう。まずは、CDでそれぞれの問題を聞きます。問題を聞いた後に一時停止ボタンを押して、それぞれの問題に対して約1分間で回答します。回答は声に出して言いましょう。

FOLLOW-UP QUESTIONS

1. How could this incident have been prevented?
2. What safety advice is given at the flight crew briefing?
3. What are some of the before-takeoff duties for the cabin crew?

UNIT 5   TAKEOFF & CLIMB

AVIATION READING

VOCABULARY

以下のキーワードを読みましょう。分からないキーワードに○をして、意味を調べましょう。

KEY WORDS

1. aft door
2. airspeed
3. airstair
4. cabin attendant
5. cabin pressure
6. cockpit
7. hijack
8. parachute
9. refuel
10. stall
11. warning light

PRONUNCIATION CD57
CDで上のキーワードを聞き、後についてリピートしましょう。何回か繰り返します。

FLUENCY CD58
CDで以下の文を聞き、音読しましょう。何回か音読を繰り返します。

AFT DOOR OPEN

The story of Northwest Orient Flight 305 is one of the most famous hijackings in the history of aviation. On November 24th 1971, a passenger called Dan Cooper boarded a Boeing 727 at Portland International Airport in the United States. After takeoff, Cooper informed a cabin attendant that he had a bomb in his briefcase. He had three demands: $200,000 in cash; four parachutes; and a fuel truck to refuel the aircraft at Seattle.

When the plane landed at Seattle-Tacoma Airport, Cooper was given the money and parachutes. He then allowed the other passengers plus two cabin attendants to get off the aircraft, leaving him and four crew still on board. The aircraft refuelled and took off again, flying southeast towards Mexico City.

Cooper ordered the captain to fly at a maximum altitude of 10,000 feet and at the lowest airspeed possible without stalling the aircraft. He also told all four crew to stay in the cockpit with the door closed. Approximately 20 minutes after takeoff, a cockpit warning light indicated that the aft airstair had been activated. A little later there was a change in cabin pressure, indicating that the aft door had been opened.

The crew tried to contact Cooper using the intercom system, but there was no response. He had manually activated the aft airstair and door, and parachuted from the plane. A huge search operation began, involving the police, FBI, Air Force and National Guard, but Cooper was never found. Today, more than four decades later, the case still remains unsolved.

UNIT 6   CLIMB

"LEVEL OFF AT YOUR DISCRETION"
EQUIPMENT FAILURE & ROUGH AIR

VOCABULARY
以下のキーワードを読みましょう。分からないキーワードに〇をして、意味を調べましょう。

KEY WORDS
1. ash cloud
2. landing area
3. level off
4. malfunction
5. non-normal situation
6. rough air
7. stall
8. terrain
9. visual check
10. volcanic eruption
11. windshield

ABBREVIATIONS
1. HDG
2. PIREP

ACRONYMS
1. CA
2. FL

What do the key words mean?

Try to explain in English!

PRONUNCIATION
CD59 CDで上のキーワードを聞き、後についてリピートしましょう。何回か繰り返します。

COMPREHENSION
CD60 CDで5つのATCの会話を聞き、それぞれの問題に答えましょう。(回答は95ページ)

LISTENING TO ATC

1. (a) What is the pilot reporting?   (b) What does the pilot think happened?
2. (a) What problem does the pilot have?   (b) What does the pilot want to do?
3. (a) What is the controller worried about?   (b) How does the pilot respond?
4. (a) Why does the pilot want to stop on the taxiway?   (b) How will the problem be fixed, do you think?
5. (a) What must the pilot do before departure?   (b) What is 'holdover time'?

FLUENCY
CD61 CDで以下の文を聞き、音読しましょう。何回か音読を繰り返します。

DESCRIBING A SINGLE PICTURE

It looks like the airplane in this picture is in the climb segment of the flight. There are some clouds around the aircraft and also some turbulence. During this segment of flight, the cabin has to be secured in case there is any rough air. The cabin crew are seated and the passengers must have their seatbelts fastened. During climb, the crew must be ready to handle any type of non-normal situation. For example, there may be birds in the area or there may be some kind of equipment failure. After the climb is completed, the passengers can move around the cabin and the CAs can begin the food or drink service.

UNIT 6 CLIMB

SINGLE PICTURE

Give enough information!

Try to be clear and accurate!

FLUENCY

絵の状況を描写しましょう。まず、絵の中から3つか4つの話題を見つけます（例えば、emergency、terrain、landing等）。次にそれぞれの話題に関して1分程度話します。目標は、3分間話すこと、完全文を使うこと、長い沈黙を避けること、er、you know、wellなどのつなぎ語を避けることです。

INTERACTIONS

CD62 上の絵について以下の問題に答えましょう。まずは、CDでそれぞれの問題を聞きます。問題を聞いた後に一時停止ボタンを押して、それぞれの問題に対して約1分間で回答します。回答は声に出して言いましょう。

FOLLOW-UP QUESTIONS

1. Name some other likely landing areas.
2. What does the pilot need to do to prepare for this type of landing?
3. What should the pilot say to the passengers?

UNIT 6 CLIMB

ATC DESCRIPTION

INTERACTIONS

CD63 下の絵について以下の問題に答えましょう。まずは、CDでそれぞれの問題を聞きます。問題を聞いた後に一時停止ボタンを押して、それぞれの問題に対して約1分間で回答します。回答は声に出して言いましょう。

WARM-UP QUESTIONS

1. Describe the situation in the picture.
2. Why is this situation dangerous?
3. How can the pilot avoid this situation?
4. What should the pilot report to the controller?

COMPREHENSION

CD64 CDでATCの会話を聞き、下線部を埋めましょう。(回答は95ページ)

ATC DIALOG

PILOT: Fukuoka Control, OB Air 1663, climbing out of 8,000 for __1._____. We have a PIREP.

CONTROLLER: OB Air 1663, Fukuoka Control, __2._____, go ahead with your PIREP.

PILOT: OB Air 1663, we encountered possible __3._____ between 5,000 and 8,000, 20 miles southeast of Mt. Sakura. Request heading change to avoid further encounter.

CONTROLLER: OB Air 1663, roger, turn right, __4._____, climb and maintain 10,000. Say condition of your aircraft.

PILOT: OB Air 1663, right turn, HDG 220, climb and maintain 10,000. Our windshield is slightly scratched and obscuring our __5._____, but all engine instruments indicate normal operation. We will continue on to our destination airport.

CONTROLLER: Roger, OB Air 1663, thank you for your __6._____.

FLUENCY

上のATCの状況を描写しましょう。1〜2分話すようにします。

INTERACTIONS

CD65 上のATCの状況について、以下の問題に答えましょう。まずは、CDでそれぞれの問題を聞きます。問題を聞いた後に一時停止ボタンを押して、それぞれの問題に対して約1分間で回答します。回答は声に出して言いましょう。

FOLLOW-UP QUESTIONS

1. How did the pilot handle this situation?
2. What should the pilot do if the engine begins to malfunction?
3. What is a PIREP? Give some examples.

UNIT 6   CLIMB

USEFUL LANGUAGE

STRUCTURE

このエクササイズでは、過去進行形と単純過去形を使って、過去の予期せぬ出来事を表わします。下の単語を並べ替えて文を作りましょう。1問目の例を参考にしてください。

INTERRUPTED ACTIONS IN THE PAST

1. failed. / was / The aircraft / the cockpit heat / when / leveling off
   → _The aircraft was leveling off when the cockpit heat failed._
2. it encountered / when / was climbing / severe turbulence. / The plane / to FL 270
   → _____
3. were / when / The cabin attendants / happened. / the bird strike / serving drinks
   → _____
4. through 10,000 feet / sounded. / when / were climbing / a cockpit warning / The pilots
   → _____

PRONUNCIATION

CD66 CDで上の文を聞き、後についてリピートしましょう。何回か繰り返します。(回答は95ページ)

STRUCTURE

下の絵は、さまざまな予期せぬ出来事を描写しています。下の「USEFUL STRUCTURE」とそれぞれの絵の上の単語を使って、起きた事を表わしましょう。1問目の例を参考にしてください。

USEFUL STRUCTURE

PAST CONTINUOUS TENSE		SIMPLE PAST TENSE
The aircraft was climbing	when	it encountered…
The pilots were leveling off		they noticed…
etc		etc

2 7,000 feet • engine • fail

1 FL 150 • ash cloud • volcanic eruption

The plane <u>was climbing</u> through FL 150 <u>when</u> the pilots <u>saw</u> the ash cloud.

The pilots <u>were climbing</u> through FL 150 <u>when</u> they <u>noticed</u> the volcanic eruption.

3 clouds • icing conditions • stall

Use past tenses for the picture sequence!

4 FL 230 • wing panel • fall off

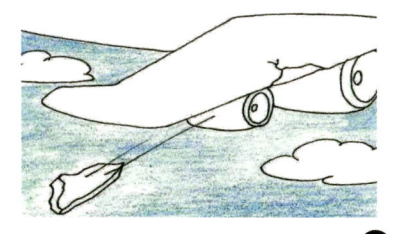

43

UNIT 6　　　　　　　　　　　　　　　　　　　　　　　　　　　CLIMB

PICTURE SEQUENCE

ADVICE　つなぎ語を避けよう

「PICTURE SEQUENCE」の問題では、話が常に流れるよう気を付けます。つなぎ語や長い沈黙、その他の意味のない音は、聞き手にとってまぎらわしく、話を理解するのを難しくしてしまいます。遅れすぎない回答が、効果的なATCコミュニケーションのポイントです。

X "There is er . . . an airplane . . . um . . . um . . . in trouble."　　**O** "There is an airplane in trouble."

FLUENCY

1～4の順に、絵の状況を描写しましょう。目標は、3分間話すこと、完全文を使うこと、長い沈黙を避けること、er、you know、wellなどのつなぎ語を避けることです。

INTERACTIONS

CD67 上の絵について以下の問題に答えましょう。まずは、CDでそれぞれの問題を聞きます。問題を聞いた後に一時停止ボタンを押して、それぞれの問題に対して約1分間で回答します。回答は声に出して言いましょう。

FOLLOW-UP QUESTIONS

1. Did the pilot make a good decision on his landing area, do you think? Why (not)?
2. How does a pilot decide a good emergency landing area?
3. How can a pilot get help during an emergency?

UNIT 6 CLIMB

AVIATION READING

VOCABULARY

以下のキーワードを読みましょう。分からないキーワードに○をして、意味を調べましょう。

KEY WORDS
1. broken ankle
2. cumulonimbus clouds
3. deviation
4. encounter
5. forecast
6. frontal system
7. galley
8. jet stream
9. moderate turbulence
10. severe turbulence
11. weather build up

PRONUNCIATION

CD68 CDで上のキーワードを聞き、後についてリピートしましょう。何回か繰り返します。

FLUENCY

CD69 CDで以下の文を聞き、音読しましょう。何回か音読を繰り返します。

FASTEN YOUR SEATBELTS!

Severe air turbulence can come from a number of sources, but it is usually associated with thunderstorms and jet streams near frontal systems. It can be difficult to avoid as it cannot be seen directly.

On April 11th 2011, American Airlines Flight 170 was climbing away from Narita Airport when the pilots noticed a small weather build up ahead of them. The captain requested a deviation in order to avoid possible turbulence, but the controller could not give a clearance right away and instead told the pilots to stand by.

A short time later the controller instructed the pilots to make a turn, but the Boeing 777 had already entered the weather. The aircraft was climbing through Flight Level 240 when it was hit by moderate and then severe turbulence. No passengers were injured because the seatbelt sign was still on. However, some of the cabin crew were working in the aft galley when the turbulence struck, and as a result four cabin attendants (CAs) were injured. Although two of the CAs had broken ankles, the captain decided to continue the flight to Los Angeles.

The pilots had not been given any forecasts of turbulence, and at the time of the incident there were no pilot reports (PIREPS) for turbulence along the flight path. The NTSB concluded that the cause of the incident was an unexpected encounter with turbulence after entering cumulonimbus clouds.

UNITS 1-6

REVIEW 1

PRONUNCIATION
「B/V」の発音に注意して、以下の文を何回か音読しましょう。

B/V IN INITIAL POSITION

1. The **b**raking action was poor due to icy conditions on the runway.
2. The pilot requested radar **v**ectors to the airport.
3. **B**irds were reported near the **v**icinity of the airport.
4. A **v**ery **b**ig thunderstorm was approaching from the north.
5. The **b**oarding **b**ridge was moved to spot twelve.
6. Poor **v**isibility and **b**lowing snow caused many flight delays.

For "V", put your top front teeth and bottom lip together, then open your mouth!

For "B", put your lips together, then open them quickly!

VOCABULARY
以下のキーワードを英語で説明してみましょう。

KEY WORDS

1. air turnback
2. anti-icing
3. baggage car
4. hazard
5. holdover time
6. lightning strike
7. overrun
8. refuel
9. rough air
10. shut down
11. tow tractor
12. turbulence
13. weather build up

What do the key words mean?

STRUCTURE
下の単語を並べ替えて文を作りましょう。その後文を何回か音読します。(回答は96ページ)

SENTENCES

1. a tow car / to / The controller / send / the runway. / has to
   → _____

2. are / the end of / Many birds / flying / the runway. / near
   → _____

3. climbed / of / conditions. / the icing / The captain / out
   → _____

4. There / a suitcase / to be / the taxiway. / seems / on
   → _____

5. Taxiway Bravo. / leaking from / onto / is / Oil / the airplane
   → _____

6. very / were / yesterday. / The runways / and slippery / wet
   → _____

7. leveling off / the wing. / lightning struck / The airplane / when / was
   → _____

8. might / Volcanic ash / during flight. / the engines / cause / to malfunction
   → _____

UNITS 1-6

REVIEW 1

PRONUNCIATION

「S/TH」の発音に注意して、以下の文を何回か音読しましょう。

S/TH IN INITIAL POSITION

1. The controller changed the runways due to the **s**trong crosswinds.
2. A big **th**understorm is approaching from the **s**outheast.
3. The pilot applied reverse **th**rust after landing.
4. The pilot encountered **s**evere turbulence at FL 2**6**0.
5. The aircraft took off from **th**ree **s**ix right.
6. I **th**ink the heavy **s**nowstorm will cause flight delays all day long.

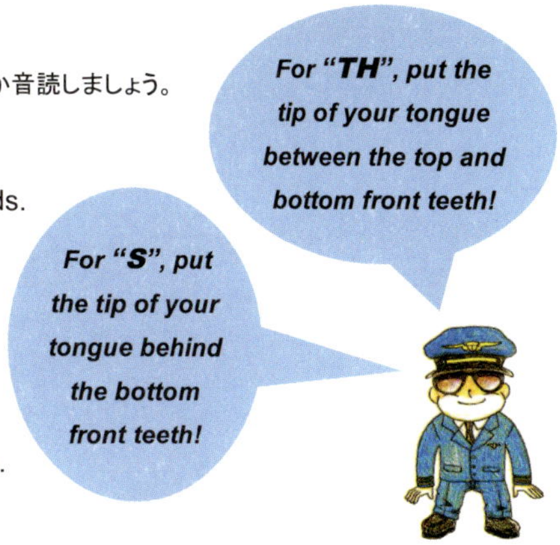

For "**TH**", put the tip of your tongue between the top and bottom front teeth!

For "**S**", put the tip of your tongue behind the bottom front teeth!

VOCABULARY

以下のキーワードを使って、下線部を埋めましょう。(回答は96ページ)

KEY WORDS

a. all areas
b. Chitose Airport
c. Gate 22
d. reduced visibility
e. re-opening
f. snowstorm

ATC DIALOG

CONTROLLER: All aircraft, __1._____ is now closed due to a __2._____. Expect further information in about 30 minutes.

PILOT: Chitose TWR, OB Air 1663, request estimated time of __3._____.

CONTROLLER: OB Air 1663, Chitose TWR, the storm is moving quickly, but we will need to remove snow from __4._____ of the airport. Time of re-opening is unknown.

PILOT: Roger, request taxi back to __5._____.

CONTROLLER: OB Air 1663, taxi back to Gate 22 at your discretion. Taxi with extreme caution due to blowing snow and __6._____.

PILOT: Roger, OB Air 1663, Gate 22.

FLUENCY

上のATCの状況について、以下の問題に答えましょう。

FOLLOW-UP QUESTIONS

1. Describe the situation in the ATC dialog.
2. How would the captain explain the situation to the passengers?
3. What are some other situations in which an airport would be closed?

UNITS 1-6

REVIEW 1

PRONUNCIATION

「L/R」の発音に注意して、以下の文を何回か音読しましょう。

L/R IN INITIAL POSITION

1. The **l**oad factor was **l**ess than sixty percent.
2. **R**udder pedals are often used during taxi and takeoff.
3. The PAPI **l**ights on **R**WY 36L were out of service.
4. The **l**anding gear was **r**etracted after takeoff.
5. The **L**earjet was **r**esponsible for the **r**unway intrusion.
6. The pilot **r**eturned to the **r**amp area because of a **l**eaking engine.

For "L", touch the tip of your tongue to the top of your mouth!

For "R", lift the middle of your tongue but don't touch the top of your mouth!

VOCABULARY

以下のキーワードを英語で説明してみましょう。

KEY WORDS

1. airstair
2. bird strike
3. broken ankle
4. debris
5. fatalities
6. ignite
7. leak
8. puncture
9. snow chains
10. snow removal
11. stuck landing gear
12. terrain
13. thunderstorm

Use simple sentences!

STRUCTURE

下の単語を並べ替えて文を作りましょう。その後文を何回か音読します。(回答は97ページ)

SENTENCES

1. is / and runways / covering / at Haneda Airport. / Snow / the taxiways
   → _____

2. is / at / some kind of / There / the ramp area. / obstruction
   → _____

3. before / The pilot / receive / taking off. / did not / his clearance
   → _____

4. when / was taxiing / began to lift. / The aircraft / to the runway / the fog
   → _____

5. an emergency landing / is / a sick passenger. / due to / The pilot / making
   → _____

6. may / if / changes. / the wind direction / change runways / The controller
   → _____

7. takeoffs / must / The passengers / and landings. / during / remain seated
   → _____

8. began. / when / was holding / The pilot / on the runway / the earthquake
   → _____

UNITS 1-6

REVIEW 1

VOCABULARY
以下の略称と頭字語を英語で説明してみましょう。

ABBREVIATIONS
1. HDG
2. PIREP
3. RWY
4. T/O
5. TWR

ACRONYMS
1. ATC
2. EICAS
3. FAA
4. FBO
5. FL
6. PAPI
7. PF
8. V_1

Try to explain in English!

VOCABULARY
以下のキーワードを使って、下線部を埋めましょう。(回答は97ページ)

KEY WORDS
a. fluid
b. main landing gear
c. microbursts
d. ramp area
e. takeoff roll
f. tow truck

ATC DIALOG

CONTROLLER: OB Air 1663, Tokyo TWR, use caution, __1._____ reported in the vicinity of the airport, wind 150 at 16, RWY 16RL, cleared for T/O.

PILOT: Tokyo TWR, OB Air 1663, cleared for T/O, RWY 16R, starting __2._____.

PILOT: Tokyo TWR, OB Air 1663, rejecting T/O, we seem to have a problem with our __3._____, we are holding on the runway.

CONTROLLER: OB Air 1663, roger, understand your situation, hold your position, we now see some kind of __4._____ leaking from your aircraft.

PILOT: OB Air 1663, we are getting an indication from our hydraulics. Request towback to __5._____.

CONTROLLER: Roger, OB Air 1663, we'll send a __6._____ to the runway.

FLUENCY
上のATCの状況について、以下の問題に答えましょう。

FOLLOW-UP QUESTIONS
1. Describe the situation in the ATC dialog.
2. What do you think happened to this flight?
3. What must be done before the runway can be used again?

UNIT 7 　　　　　　　　　　　　　　　　　　　　　　　CRUISE

"IS THERE A DOCTOR ON BOARD?"
PASSENGER INJURIES & PROBLEMS

VOCABULARY
以下のキーワードを読みましょう。分からないキーワードに〇をして、意味を調べましょう。

KEY WORDS
1. chief purser
2. destination
3. diversion
4. divert
5. drunk
6. evacuation
7. medical assistance
8. medicine
9. police assistance
10. pregnant
11. restrain
12. stomach ache
13. unruly behavior

ACRONYM
1. ETA

PRONUNCIATION
CD70 CDで上のキーワードを聞き、後についてリピートしましょう。何回か繰り返します。

COMPREHENSION
CD71 CDでそれぞれの対の単語を聞き、聞こえた方に〇をつけましょう。（回答は98ページ）

B/V IN INITIAL POSITION
1. **b**an – **v**an
2. **b**at – **v**at
3. **b**eer – **v**eer
4. **b**ent – **v**ent
5. **b**erry – **v**ery
6. **b**est – **v**est
7. **b**et – **v**et
8. **b**oat – **v**ote
9. **b**olt – **v**olt
10. **b**owl – **v**ole

For "B", put your lips together, then open them quickly!

PRONUNCIATION
CD72 CDで以下の表現を聞き、後についてリピートしましょう。

B/V IN VARIOUS POSITIONS
1. **b**ird activity
2. pri**v**ate pilot
3. a**b**ort takeoff
4. ATC ser**v**ices
5. esta**b**lish approach
6. flight le**v**el 150
7. fire pre**v**ention
8. **v**isibility o**b**scured
9. weather ad**v**isory
10. ca**b**in pro**b**lem
11. flight de**v**iation
12. **b**roken seat**b**elt
13. flight maneu**v**er

For "V", put your top front teeth and bottom lip together, then open your mouth!

FLUENCY
CD73 CDで以下の文を聞き、音読しましょう。何回か音読を繰り返します。

DESCRIBING A SINGLE PICTURE

It looks like there is a problem with one of the passengers. A boy seems to be sick. He is holding his stomach and crying, so I think that he might have a stomach ache. His mother is sitting next to him and she looks worried. She is trying to comfort her son. There is also a cabin attendant who is talking with the mother. Maybe the CA will offer some medicine to the child. However, if the pain gets worse the chief purser should make a doctor call, or contact the pilot and ask for further advice.

UNIT 7   CRUISE

SINGLE PICTURE

> **ADVICE** 正確な単語を選択しよう
>
> 「SINGLE PICTURE」の問題では、状況を描写する時に正確に単語を選択することが重要です。正しい単語を思い起こせない場合は、意味の近い単語や言い回しを使います。単語の選択が正確でなかったり、日本語を使ってしまうと、聞き手が混乱したり意味を取り違えてしまう可能性がありますから、注意が必要です。
>
> **X** "She has fat stomach. She is *ninshin, ninshin*!"   **O** "She has a baby in her stomach."

FLUENCY 絵の状況を描写しましょう。まず、絵の中から3つか4つの話題を見つけます（例えば、passenger problem、CA、pilot options等）。次にそれぞれの話題に関して1分程度話します。目標は、3分間話すこと、完全文を使うこと、長い沈黙を避けること、er、wellなどのつなぎ語を避けることです。

INTERACTIONS **CD74** 上の絵について以下の問題に答えましょう。まずは、CDでそれぞれの問題を聞きます。問題を聞いた後に一時停止ボタンを押して、それぞれの問題に対して約1分間で回答します。回答は声に出して言いましょう。

FOLLOW-UP QUESTIONS

1. What action would you take in this situation?
2. Should the passenger be moved? Why (not)?
3. Describe some other passenger problems.

UNIT 7   CRUISE

ATC DESCRIPTION

INTERACTIONS

CD75 下の絵について以下の問題に答えましょう。まずは、CDでそれぞれの問題を聞きます。問題を聞いた後に一時停止ボタンを押して、それぞれの問題に対して約1分間で回答します。回答は声に出して言いましょう。

WARM-UP QUESTIONS

1. Describe the situation in the picture.
2. What should the cabin crew do?
3. What can the cockpit crew do in this situation?
4. Give examples of other types of unruly behavior.

COMPREHENSION CD76 CDでATCの会話を聞き、正しい答えを選びましょう。(回答は98ページ)

ATC DIALOG

1. The passenger is…
   (a) sleeping   (b) very old   (c) drunk   (d) travelling with friends
2. The pilot wants to…
   (a) change the gate (b) calm down (c) make a diversion (d) continue to the destination
3. The pilot does not want…
   (a) police assistance (b) medical assistance (c) to taxi to the gate (d) to stop on the runway

FLUENCY 上のATCの状況を描写しましょう。1〜2分話すようにします。

INTERACTIONS CD77 上のATCの状況について、以下の問題に答えましょう。まずは、CDでそれぞれの問題を聞きます。問題を聞いた後に一時停止ボタンを押して、それぞれの問題に対して約1分間で回答します。回答は声に出して言いましょう。

FOLLOW-UP QUESTIONS

1. What did the pilot request to the controller?
2. How did the pilot handle this situation?
3. What do you think happened after the aircraft landed?

UNIT 7   CRUISE

USEFUL LANGUAGE

STRUCTURE

このエクササイズでは、助動詞の「should」を使って、問題解決のアドバイスを表わします。下の単語を並べ替えて文を作りましょう。1問目の例を参考にしてください。

SOLVING PROBLEMS

1. make / purser / a doctor call. / should / The → _The purser should make a doctor call._
2. order. / The captain / give / should / the evacuation → _____
3. fasten / Passengers / their / should / seatbelts. → _____
4. should / alcohol. / drink / not / The drunk man / more → _____
5. the unruly / restrain / The crew / passenger. / should → _____
6. should / police / The pilots / assistance. / request → _____
7. woman. / not / The CAs / move / should / the pregnant → _____
8. medicine. / should / CA / some / bring / The → _____

PRONUNCIATION

CD78 CDで上の文を聞き、後についてリピートしましょう。何回か繰り返します。（回答は98ページ）

STRUCTURE

下の絵は、乗客の問題行動を描写しています。下の「USEFUL STRUCTURE」とそれぞれの絵の上の単語を使って、解決策を表わす文を作りましょう。1問目の例を参考にしてください。

USEFUL STRUCTURE

SUBJECT	MODAL	VERB
The CA	should	request..
Cabin crew	should not	gather…
Passengers		use…
etc		etc

2 loud music • computer • earphones

3 lavatory • smoke • alarm

1 drinking • disturb • be quiet

Practise making simple, accurate sentences!

4 gather • emergency exit • seats

This passenger <u>should</u> stop drinking alcohol.
He <u>should not disturb</u> other people.
The chief purser <u>should ask</u> him to be quiet.

UNIT 7　　　　　　　　　　　　　　　　　　　　　　　　　　　　　CRUISE

PICTURE SEQUENCE

Make accurate word choices!

Avoid using fillers and reduce your pauses!

FLUENCY

1〜4の順に、絵の状況を描写しましょう。
目標は、3分間話すこと、完全文を使うこと、
長い沈黙を避けること、er、you know、wellなどのつなぎ語を避けることです。

1

2

3

4

INTERACTIONS

CD79 上の絵について以下の問題に答えましょう。まずは、CDでそれぞれの問題を聞きます。
問題を聞いた後に一時停止ボタンを押して、それぞれの問題に対して約1分間で回答します。回答
は声に出して言いましょう。

FOLLOW-UP QUESTIONS

1. How did the flight crew get help?
2. Name some factors that the pilots should consider in this situation.
3. If the pilots decided to divert, what preparations should they consider?

UNIT 7　　　　　　　　　　　　　　　　　　　　　　　　　　　　　　CRUISE

AVIATION READING

VOCABULARY

以下のキーワードを読みましょう。分からないキーワードに○をして、意味を調べましょう。

KEY WORDS
1. bruise
2. broken bone
3. carry-on bag
4. concussion
5. estimate
6. fasten seatbelt
7. overhead baggage compartment
8. phase of flight

PRONUNCIATION

CD80 CDで上のキーワードを聞き、後についてリピートしましょう。何回か繰り返します。

FLUENCY

CD81 CDで以下の文を聞き、音読しましょう。何回か音読を繰り返します。

INJURIES TO PASSENGERS

Turbulence occurs on almost every airline flight, but there is usually little danger to the passengers and crew. However, in some cases turbulence caused by weather conditions such as thunderstorms can be severe and may cause damage to the aircraft or injury to the passengers.

It is important to realize that turbulence can happen in any phase of flight, but it can be especially dangerous during the cruise phase because that is the time when passengers and crew are likely to be out of their seats and walking around the cabin.

During rough air conditions, passengers may be injured by objects falling from the overhead baggage compartments. If the compartments are not securely locked, they may open and carry-on bags might fall out. A study by the Flight Safety Foundation estimated that about 10,000 passengers are injured every year from falling baggage. These injuries range from cuts and bruises to broken bones and severe head injuries like concussions. In rare cases, falling objects may cause fatalities.

What can the flight crew do to make the flight safe and comfortable for passengers? Firstly, when the captain turns on the 'Fasten Seatbelts' sign, the crew should make sure that passengers return to their seats and fasten their seatbelts. Secondly, passengers should always wear their seatbelts when seated, even if the flight is smooth. Finally, passengers should always take care when opening overhead baggage compartments.

UNIT 8 — EMERGENCY

"MAYDAY, MAYDAY, MAYDAY!"
SMOKE IN THE CABIN

VOCABULARY

以下のキーワードを読みましょう。分からないキーワードに○をして、意味を調べましょう。

KEY WORDS

1. alternate airport
2. controlled descent
3. cruise flight
4. emergency procedure
5. fire extinguisher
6. flight plan
7. oxygen mask
8. priority landing
9. radar vectors
10. rapid decompression
11. warning

PRONUNCIATION

CD82 CDで上のキーワードを聞き、後についてリピートしましょう。何回か繰り返します。

COMPREHENSION

CD83 CDでそれぞれの対の単語を聞き、聞こえた方に○をつけましょう。(回答は99ページ)

S/TH IN INITIAL POSITION

1. **s**ank – **th**ank
2. **s**aw – **th**aw
3. **s**ick – **th**ick
4. **s**igh – **th**igh
5. **s**in – **th**in
6. **s**ing – **th**ing
7. **s**ink – **th**ink
8. **s**ome – **th**umb
9. **s**ong – **th**ong
10. **s**ort – **th**ought

For "S", put the tip of your tongue behind the bottom front teeth!

PRONUNCIATION

CD84 CDで以下の表現を聞き、後についてリピートしましょう。

S/TH IN VARIOUS POSITIONS

1. de**s**tination
2. nor**th**we**s**t
3. medical a**ss**i**s**tance
4. oxygen ma**s**k
5. chief pur**s**er
6. **s**ou**th**ea**s**t
7. police a**ss**i**s**tance
8. rever**s**e **th**ru**s**t
9. **s**trong cro**ss**wind
10. fir**s**t cla**ss** pa**ss**enger
11. civil aviation au**th**ority
12. fa**s**ten **s**eatbelt**s**
13. indicated air**s**peed

For "TH", put the tip of your tongue between the top and bottom front teeth!

FLUENCY

CD85 CDで以下の文を聞き、音読しましょう。何回か音読を繰り返します。

DESCRIBING A SINGLE PICTURE

It looks like this airplane is in cruise flight, but there is an emergency situation. The left side door has broken off from the forward section of the cabin. This will cause a rapid decompression. I think the pilots have already received a warning in the flight deck. They must make a controlled descent to stabilize the pressure in the cabin. The cabin attendants must do the emergency procedure. After the descent they will make sure that the passengers are calm, and check that they have their oxygen masks on correctly.

UNIT 8 EMERGENCY

SINGLE PICTURE

Don't use Japanese words!   *Make accurate word choices!*

FLUENCY

絵の状況を描写しましょう。まず、絵の中から3つか4つの話題を見つけます（例えば、cabin、unruly passenger、CA等）。次にそれぞれの話題に関して1分程度話します。目標は、3分間話すこと、完全文を使うこと、長い沈黙を避けること、er、wellなどのつなぎ語を避けることです。

INTERACTIONS

CD86 上の絵について以下の問題に答えましょう。まずは、CDでそれぞれの問題を聞きます。問題を聞いた後に一時停止ボタンを押して、それぞれの問題に対して約1分間で回答します。回答は声に出して言いましょう。

FOLLOW-UP QUESTIONS

1. What are some reasons for a pilot to change the destination during cruise flight?
2. Name some types of equipment failure which could make pilots change their destination.
3. What does a pilot need to do to change the flight plan in an emergency situation?

UNIT 8　　　　　　　　　　　　　　　　　　　　　　　　　　EMERGENCY

ATC DESCRIPTION

INTERACTIONS

CD87 下の絵について以下の問題に答えましょう。まずは、CDでそれぞれの問題を聞きます。問題を聞いた後に一時停止ボタンを押して、それぞれの問題に対して約1分間で回答します。回答は声に出して言いましょう。

WARM-UP QUESTIONS

1. Describe the situation in the picture.
2. Why are the pilots wearing masks?
3. What is '7700'?
4. How can the pilots communicate with each other and the ATC controller?

COMPREHENSION

CD88 CDでATCの会話を聞きながら3つの間違いに○をして、正しい言葉を書きましょう。
（回答は99ページ）

ATC DIALOG

PILOT:　　　　　Tokyo Control, OB Air 1663, we are making an emergency descent to a lower altitude due to rapid decompression in the cabin. The oxygen masks have been deployed. Request radar vectors to the nearest airport.

CONTROLLER:　　OB Air 1663, Tokyo Control, roger, understand your situation, turn right, heading 280, descend and maintain 10,000. We will give you radar vectors to Chuubu Airport. Please say the condition of your ship and passengers.

PILOT:　　　　　Descending to 10,000, all passengers are accounted for, no injuries reported. We do not know the extent of the damage yet, but there seems to be some problem near the overhead baggage compartment in the forward section of the cabin.

FLUENCY

上のATCの状況を描写しましょう。1～2分話すようにします。

INTERACTIONS

CD89 上のATCの状況について、以下の問題に答えましょう。まずは、CDでそれぞれの問題を聞きます。問題を聞いた後に一時停止ボタンを押して、それぞれの問題に対して約1分間で回答します。回答は声に出して言いましょう。

FOLLOW-UP QUESTIONS

1. How should the cockpit and cabin crew coordinate their actions during this situation?
2. Describe some emergencies that could occur during cruise flight.
3. Name some situations that would cause an emergency descent.

UNIT 8　　　　　　　　　　　　　　　　　　　　　　　　　　　　　EMERGENCY

USEFUL LANGUAGE

STRUCTURE

このエクササイズでは、「will」と「going to」を使って、未来の出来事を表わします。
下の単語を並べ替えて文を作りましょう。1問目の例を参考にしてください。

ACTIONS IN THE FUTURE

1. will / assistance. / request / The captain / police → *The captain will request police assistance.*
2. going / The pilot / to / a controlled descent. / is / make → _____
3. unruly / restrain / passenger. / The crew / the / will → _____
4. is / to / the lavatory. / going / The purser / check → _____
5. The pilots / to / a priority landing. / are / request / going → _____
6. put / masks. / will / oxygen / The cabin crew / on → _____
7. the / will / order. / The captain / evacuation / give → _____
8. are / the fire extinguishers. / to / going / The CAs / use → _____

PRONUNCIATION

CD90 CDで上の文を聞き、後についてリピートしましょう。何回か繰り返します。(回答は99ページ)

STRUCTURE

下の絵は、フライト中のさまざまな問題を描写しています。下の「USEFUL STRUCTURE」とそれぞれの絵の上の単語を使って、この先に起こる事を表わす文を作りましょう。1問目の例を参考にしてください。

USEFUL STRUCTURE

SUBJECT		VERB
The pilots	will	notify…
The CA	is going to	check…
The passengers	are going to	request…
etc		etc

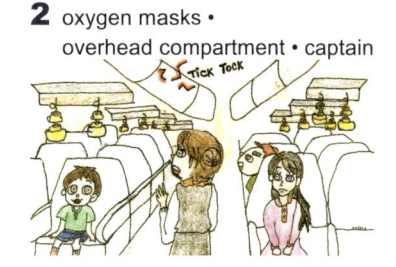

2 oxygen masks • overhead compartment • captain

1 emergency descent • divert • medical assistance

Use complete sentences!

3 emergency • fire • oxygen masks

4 run out • priority landing • alternate airport

The pilots <u>are going to</u> make an emergency descent.
They <u>will</u> divert to the nearest airport.
They <u>are going to</u> request medical assistance.

UNIT 8   EMERGENCY

PICTURE SEQUENCE

ADVICE 明瞭かつ率直に

緊急事態では、明確で直接的な指示を出さなければなりません。発音も明瞭かつ正確を心がけます。そうすることで、回答は理解されやすくなるでしょう。

X "I see smoke . . . I see fire . . . erhh . . . I see cabin attendant." **O** "The cabin attendant found a fire in the galley."

FLUENCY

1～4の順に、絵の状況を描写しましょう。目標は、3分間話すこと、完全文を使うこと、長い沈黙を避けること、er、you know、wellなどのつなぎ語を避けることです。

INTERACTIONS

CD91 上の絵について以下の問題に答えましょう。まずは、CDでそれぞれの問題を聞きます。問題を聞いた後に一時停止ボタンを押して、それぞれの問題に対して約1分間で回答します。回答は声に出して言いましょう。

FOLLOW-UP QUESTIONS

1. What do you think happened next?
2. In this situation, what were the cockpit crew's duties after landing?
3. Name some important points when handling an emergency situation.

UNIT 8   EMERGENCY

AVIATION READING

VOCABULARY
以下のキーワードを読みましょう。分からないキーワードに○をして、意味を調べましょう。

KEY WORDS
1. emergency landing
2. flame out
3. fuel supply
4. glide
5. leading edge
6. rate of descent
7. St Elmo's fire
8. transponder code
9. water ditching
10. weather radar
11. windscreen

PRONUNCIATION
CD92 CDで上のキーワードを聞き、後についてリピートしましょう。何回か繰り返します。

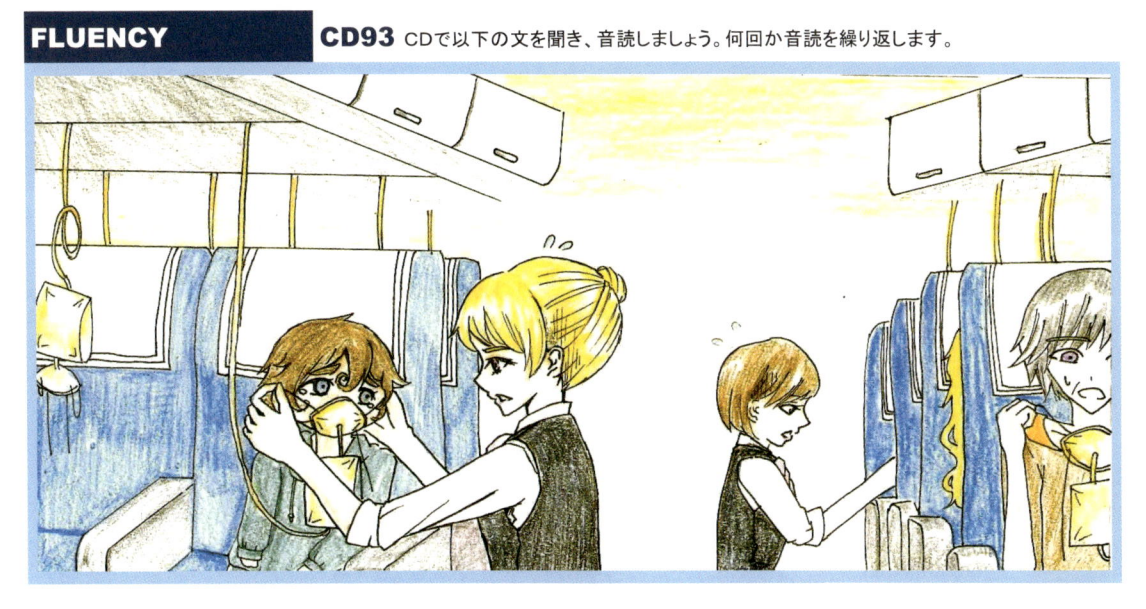

FLUENCY
CD93 CDで以下の文を聞き、音読しましょう。何回か音読を繰り返します。

SMOKE IN THE CABIN

On June 24th 1982, British Airways Flight 9 was in cruise flight above the Indian Ocean when it encountered a volcanic ash cloud. The flight crew first noticed a strange light like St Elmo's fire on the windscreen and the leading edge of the wings. As a precaution, they switched on the engine anti-icing and passenger seatbelt signs.

Smoke started to enter the cabin, and soon after one of the engines surged and flamed out. The flight crew performed the engine shutdown procedure to shut off the fuel supply and arm the fire extinguishers. However, within minutes the other three engines of the Boeing 747 had all flamed out. The captain declared an emergency and squawked the emergency transponder code '7700'.

With all four engines shut down, the aircraft was steadily losing altitude. The crew had to decide whether to attempt a water ditching in the ocean or try to glide over mountains to the airport at Jakarta, in Indonesia. After repeatedly trying to restart the engines, they finally succeeded as the plane descended through 13,000 feet. The crew were then able to slow the rate of descent, and make a three-engine emergency landing at Jakarta. The windscreen had been badly scratched by volcanic ash so they had to make the approach on instruments.

It was later found that Flight 9 had flown through an ash cloud from the eruption of Mount Galunggung, on the island of Java. The ash was dry and therefore did not show up on the aircraft's weather radar, which is designed to detect moisture in clouds. The ash sandblasted the windscreen, and caused so much damage to the engines that three replacement engines had to be installed before the aircraft could fly again.

UNIT 9   HOLDING

"THE AIRPORT IS NOW CLOSED"
BAD WEATHER & NATURAL DISASTERS

VOCABULARY
以下のキーワードを読みましょう。分からないキーワードに〇をして、意味を調べましょう。

KEY WORDS
1. aircraft status
2. climb and maintain
3. fuel status
4. hold time
5. indefinitely
6. oval-shaped
7. racetrack
8. re-open
9. runway sweep
10. stack
11. standby power

PRONUNCIATION
CD94 CDで上のキーワードを聞き、後についてリピートしましょう。何回か繰り返します。

COMPREHENSION
CD95 CDでそれぞれの対の単語を聞き、聞こえた方に〇をつけましょう。（回答は100ページ）

L/R IN INITIAL POSITION
1. lack – rack
2. lake – rake
3. lamp – ramp
4. land – rand
5. lane – rain
6. late – rate
7. lead – read
8. leak – reek
9. lift – rift
10. light – right
11. load – road
12. long – wrong
13. low – row

For "**L**", touch the tip of your tongue to the top of your mouth!

PRONUNCIATION
CD96 CDで以下の表現を聞き、後についてリピートしましょう。

L/R IN VARIOUS POSITIONS
1. IFR clearance
2. ferry flight
3. meteorology
4. flight operations
5. reciprocating engine
6. low level turbulence
7. cruise flight
8. aeronautical chart
9. flight plan
10. landing area
11. controlled descent
12. glide slope
13. hydraulic fluid

For "**R**", lift the middle of your tongue but don't touch the top of your mouth!

FLUENCY
CD97 CDで以下の文を聞き、音読しましょう。何回か音読を繰り返します。

DESCRIBING A SINGLE PICTURE

I can see five aircraft in a holding pattern. Three of the aircraft are in the stack, one is entering and one is leaving. The holding pattern is an oval-shaped racetrack. All the planes are flying in the same direction, but they are flying at different altitudes. The airplanes are waiting for the controller to give them clearance to land, and the plane at the bottom will land first. There are many reasons why aircraft must wait before landing. For example, the airport may be very busy now or the weather may be too bad to land safely. Also, a runway sweep could be in progress.

UNIT 9   HOLDING

SINGLE PICTURE

ADVICE 確認しましょう

「SINGLE PICTURE」の問題で、質問の意味や状況の理解が難しい時には、確認しましょう。管制官にさらに情報をリクエストしたり「say again」と願い出ます。ただ「I don't know!」と回答するのは避けましょう。

X "I don't know!"   **O** "I don't know this weather hazard. I should contact the controller and get confirmation."

FLUENCY

絵の状況を描写しましょう。まず、絵の中から3つか4つの話題を見つけます（例えば、weather、holding、pilot options等）。次にそれぞれの話題に関して1分程度話します。目標は、3分間話すこと、完全文を使うこと、長い沈黙を避けること、er、wellなどのつなぎ語を避けることです。

INTERACTIONS

CD98 上の絵について以下の問題に答えましょう。まずは、CDでそれぞれの問題を聞きます。問題を聞いた後に一時停止ボタンを押して、それぞれの問題に対して約1分間で回答します。回答は声に出して言いましょう。

FOLLOW-UP QUESTIONS

1. How will the weather in this picture affect arrivals and departures at the airport?
2. What decisions do the pilots need to make if the airport is closed indefinitely?
3. What other situations would cause an airport to be completely closed?

UNIT 9 — HOLDING

ATC DESCRIPTION

INTERACTIONS

CD99 下の絵について以下の問題に答えましょう。まずは、CDでそれぞれの問題を聞きます。問題を聞いた後に一時停止ボタンを押して、それぞれの問題に対して約1分間で回答します。回答は声に出して言いましょう。

WARM-UP QUESTIONS

1. Describe the situation in the picture.
2. What is the cause of this situation?
3. How will the ATC controller handle this problem?
4. Have you experienced a runway or airport being closed? If so, describe what happened.

COMPREHENSION

CD100 CDでATCの会話を聞き、下線部を埋めましょう。(回答は100ページ)

ATC DIALOG

CONTROLLER: All aircraft, Haneda Airport is now closed due to a major __1._____. We need to do a runway sweep of all runways before __2._____. Stand by for further. BREAK BREAK. OB Air 1663, Tokyo Tower, go around, turn right heading 100, climb and maintain 7,000, hold over CHIBA.

PILOT: OB Air 1663, roger, 7,000, right heading 100, be advised, we can only hold 60 minutes, request condition of Narita Airport for __3._____.

CONTROLLER: OB Air 1663, Narita Airport is now closed. Narita is on __4._____ and damage reported to the tower facility. Re-open __5._____ is unknown at this time.

PILOT: OB Air 1663, roger, we'll hold until we get more information from the company.

CONTROLLER: Understand, OB Air 1663, we'll keep you updated on the situation, but it appears to have been a pretty big earthquake. A __6._____ has been issued at Haneda Airport.

FLUENCY

上のATCの状況を描写しましょう。1〜2分話すようにします。

INTERACTIONS

CD101 上のATCの状況について、以下の問題に答えましょう。まずは、CDでそれぞれの問題を聞きます。問題を聞いた後に一時停止ボタンを押して、それぞれの問題に対して約1分間で回答します。回答は声に出して言いましょう。

FOLLOW-UP QUESTIONS

1. How would you handle this situation?
2. How does an airport prepare for a situation like this?
3. Have you ever had a similar experience? If so, describe what happened.

UNIT 9 HOLDING

USEFUL LANGUAGE

STRUCTURE

このエクササイズでは、「want to」と「need to」の表現を使って、さまざまな人達がしたい事やする必要がある事を表わします。下の単語を並べ替えて文を作りましょう。1問目の例を参考にしてください。

WANTS & NEEDS

1. get / wants / information. / The pilot / more / to → _The pilot wants to get more information._
2. The CA / the emergency / check / to / exit. / needs → ____
3. to / wants / assistance. / request / The captain / police → ____
4. for / The pilots / clearance. / wait / need / to → ____
5. ETA. / wants / The controller / to / the / know → ____
6. solve / want / quickly. / the problem / They / to → ____
7. to / to / The passenger / move / another seat. / wants → ____
8. to / need / The pilots / divert / the alternate airport. / to → ____

PRONUNCIATION

CD102 CDで上の文を聞き、後についてリピートしましょう。何回か繰り返します。（回答は100ページ）

STRUCTURE

下の絵は、さまざまな緊急事態を描写しています。下の「USEFUL STRUCTURE」とそれぞれの絵の上の単語を使って、状況を表わす文を作りましょう。1問目の例を参考にしてください。

USEFUL STRUCTURE

SUBJECT		VERB
The pilots	want to	know…
The controller	wants to	find out…
The police	need to	request…
etc	needs to	etc

Use the structure to help you speak in complete sentences!

1 damage report • wait time • intentions

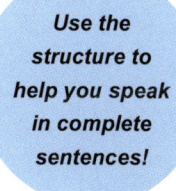

2 system problem • hold time • priority landing

3 hijacker • demands • aircraft status

4 hold time • fuel status • re-open time

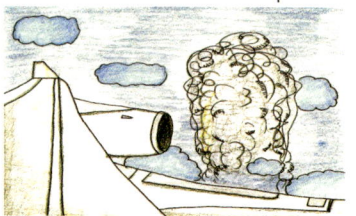

The pilots <u>want to</u> know the damage report.
Also, they <u>need to</u> find out the wait time.
The controller <u>wants to</u> know the pilot's intentions.

65

UNIT 9   HOLDING

PICTURE SEQUENCE

Speak clearly!   *Be direct and accurate!*

FLUENCY

1～4の順に、絵の状況を描写しましょう。
目標は、3分間話すこと、完全文を使うこと、
長い沈黙を避けること、er、you know、wellなどのつなぎ語を避けることです。

INTERACTIONS

CD103 上の絵について以下の問題に答えましょう。まずは、CDでそれぞれの問題を聞きます。問題を聞いた後に一時停止ボタンを押して、それぞれの問題に対して約1分間で回答します。回答は声に出して言いましょう。

FOLLOW-UP QUESTIONS

1. What had to be done before this airport could become operational again?
2. If no alternate airports were available nearby, what options did the pilots have?
3. What are some factors that pilots must consider when choosing a suitable alternate airport?

UNIT 9   HOLDING

AVIATION READING

VOCABULARY

以下のキーワードを読みましょう。分からないキーワードに○をして、意味を調べましょう。

KEY WORDS

1. autopilot
2. congestion
3. fatigue
4. flight director
5. fuel exhaustion
6. fuel load
7. maintenance problem
8. missed approach
9. stress
10. windshear
11. workload

PRONUNCIATION

CD104 CDで上のキーワードを聞き、後についてリピートしましょう。何回か繰り返します。

FLUENCY

CD105 CDで以下の文を聞き、音読しましょう。何回か音読を繰り返します。

FUEL EXHAUSTION

On January 25th 1990, Avianca Flight 052 was flying from Columbia to John F. Kennedy International Airport in New York when it encountered poor weather in the northeastern part of the United States. As a result of the weather and congestion, the aircraft had to enter three holding patterns for a total of almost 80 minutes.

During the third holding period, the flight crew notified ATC that they could only hold for about five more minutes because they were running out of fuel. Finally, the Boeing 707 was cleared to descend towards JFK Airport but it then encountered windshear and the crew executed a missed approach. They were trying to return for a second approach when all four engines suffered a loss of power due to fuel exhaustion. The aircraft crashed at Cove Neck, Long Island, and 73 of the passengers and crew were killed.

The NTSB investigation concluded that the probable cause of the accident was that the crew had failed to manage their fuel load properly, and had failed to notify ATC of an emergency fuel situation. However, the report also listed several other factors that contributed to the accident. One of these factors was the windshear that prevented the aircraft from successfully completing the first approach. In addition, the flight crew were suffering from high workload, stress and fatigue. Indeed, maintenance problems with the autopilot and flight director indicated that the aircraft might have been flown manually all the way from Columbia to New York.

UNIT 10　　　　　　　　　　　　　　　　　　　　　　　　　APPROACH

"DOWN AND LOCKED?"
PROBLEMS DURING APPROACH

VOCABULARY
以下のキーワードを読みましょう。分からないキーワードに〇をして、意味を調べましょう。

KEY WORDS
1. approach speed
2. ceiling
3. final approach
4. flat tire
5. glide path
6. glide slope
7. low pass
8. noise abatement
9. residential area
10. runway incursion
11. runway number
12. visibility
13. wind check

ABBREVIATION
1. CAT 1

ACRONYM
1. RVR

What do the key words mean?

Try to explain in English!

PRONUNCIATION　CD106　CDで上のキーワードを聞き、後についてリピートしましょう。何回か繰り返します。

COMPREHENSION　CD107　CDで5つのATCの会話を聞き、それぞれの問題に答えましょう。（回答は101ページ）

LISTENING TO ATC

1. (a) What are the wind conditions for RWY 16?　　(b) What do you think the pilot should do?
2. (a) What seems to be the problem for Citation 16C?　　(b) What do the pilots want the tower to do?
3. (a) Describe the approach for RWY 16.　　(b) What does 'noise abatement' mean?
4. (a) What are the tower instructions?　　(b) What is the reason for the go around?
5. (a) What is Citation 16C going to do?　　(b) Why can't the pilots land immediately?

FLUENCY　CD108　CDで以下の文を聞き、音読しましょう。何回か音読を繰り返します。

DESCRIBING A SINGLE PICTURE

This airplane is making a turning approach. It is going to land at a single runway airport. I cannot see the runway numbers, but the approach lights on both sides of the runway show that the aircraft is on the glide slope. It looks like it might be a difficult approach because there is a residential area at the approach end of the runway. Also, the opposite end of the runway and sides are surrounded by water. I think that this airport might have noise abatement procedures because of the houses.

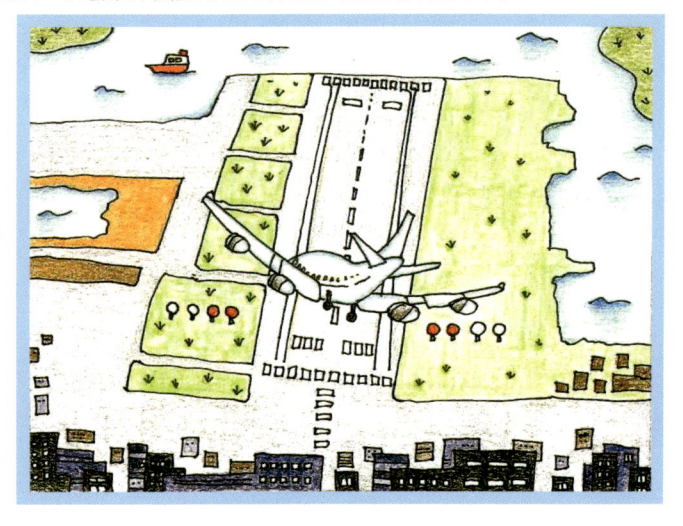

UNIT 10   APPROACH

SINGLE PICTURE

Don't use Japanese words!

Remember to use present tenses for the single picture!

FLUENCY

絵の状況を描写しましょう。まず、絵の中から3つか4つの話題を見つけます（例えば、airport、terrain、approach等）。次にそれぞれの話題に関して1分程度話します。目標は、3分間話すこと、完全文を使うこと、長い沈黙を避けること、er、wellなどのつなぎ語を避けることです。

INTERACTIONS

CD109 上の絵について以下の問題に答えましょう。まずは、CDでそれぞれの問題を聞きます。問題を聞いた後に一時停止ボタンを押して、それぞれの問題に対して約1分間で回答します。回答は声に出して言いましょう。

FOLLOW-UP QUESTIONS

1. What obstacles can you see in making this approach?
2. How may the weather conditions affect the approach?
3. Is the pilot on the proper glide path? How do you know?

UNIT 10 APPROACH

ATC DESCRIPTION

INTERACTIONS

CD110 下の絵について以下の問題に答えましょう。まずは、CDでそれぞれの問題を聞きます。問題を聞いた後に一時停止ボタンを押して、それぞれの問題に対して約1分間で回答します。回答は声に出して言いましょう。

WARM-UP QUESTIONS

1. Describe the situation in the picture.
2. What is the airplane doing?
3. What are the weather conditions at the airport?
4. Describe the procedure for a missed approach.

COMPREHENSION

CD111 CDでATCの会話を聞き、正しい答えを選びましょう。(回答は101ページ)

ATC DIALOG

1. OB Air 1663 is first instructed to…
   (a) land on RWY 01R   (b) go around RWY 01R   (c) land on RWY 01L   (d) confirm runway in sight
2. The pilot is worried about…
   (a) traffic on final   (b) his landing clearance   (c) runway visibility   (d) an obstacle on the runway
3. The pilot finally decides to…
   (a) land on RWY 01R   (b) maintain final approach   (c) hold on RWY 01L   (d) execute a missed approach

FLUENCY

上のATCの状況を描写しましょう。1～2分話すようにします。

INTERACTIONS

CD112 上のATCの状況について、以下の問題に答えましょう。まずは、CDでそれぞれの問題を聞きます。問題を聞いた後に一時停止ボタンを押して、それぞれの問題に対して約1分間で回答します。回答は声に出して言いましょう。

FOLLOW-UP QUESTIONS

1. How did the pilot handle this situation?
2. What do you think the pilot told his cabin crew and passengers?
3. What might happen if the pilot tried to land under these conditions?

UNIT 10　　　　　　　　　　　　　　　　　　　　　　　　　　　　　　APPROACH

USEFUL LANGUAGE

STRUCTURE

このエクササイズでは、「if」と「should」を使って、さまざまな状況に対するアドバイスを表わします。下の単語を並べ替えて文を作りましょう。1問目の例を参考にしてください。

CONDITIONAL ADVICE

1. go around. / should / If there are / the pilot / on the runway, / birds
   → _If there are birds on the runway, the pilot should go around._

2. should / is / If the approach speed / reduce speed. / the PF / too fast,
   → _____

3. on approach, / should not / windshear / the pilot / try to land. / If there is
   → _____

4. should / If the approach / the pilot / increase speed and pitch. / too low, / is
   → _____

PRONUNCIATION

CD113 CDで上の文を聞き、後についてリピートしましょう。何回か繰り返します。(回答は101ページ)

STRUCTURE

下の絵は、さまざまな進入状況を描写しています。下の「USEFUL STRUCTURE」とそれぞれの絵の上の単語を使って、状況に対するアドバイスを表わす文を作りましょう。1問目の例を参考にしてください。

USEFUL STRUCTURE

	CONDITION	ADVICE
If	the ceiling is too low,	the pilot should…
	there is a crosswind,	the PF should not…
	etc	etc

2 PAPI lights • landing gear • runway incursion

1 visibility • ceiling • crosswind

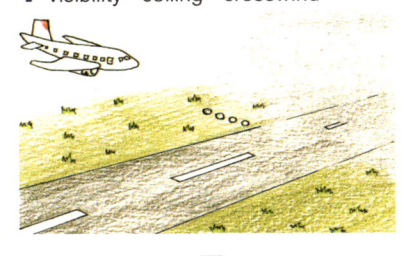

Always use complete sentences!

3 residential area • noise abatement • water

4 approach speed • glide slope • alignment

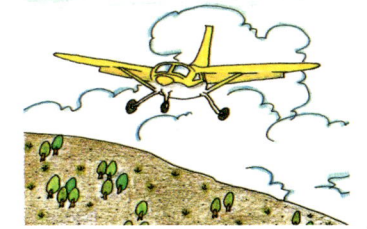

_If_ the visibility is too poor, the pilot _should_ go around.
_If_ the ceiling is too low, the pilot _should not_ try to land.
_If_ there is a crosswind, the pilot _should_ request a wind check.

UNIT 10   APPROACH

PICTURE SEQUENCE

ADVICE 文法をコントロールしよう

「PICTURE SEQUENCE」の問題では、能力に見合ったレベルで話しましょう。複雑な文法が苦手ならば、短くシンプルな文を使います。そうすれば、文法ミスを減らせるために、より正確な文になり、話や状況を理解してもらいやすくなります。

X "Something is wrong and a pilot doesn't know."   O "The airplane had a problem. The pilot didn't understand the problem."

FLUENCY

1〜4の順に、絵の状況を描写しましょう。目標は、3分間話すこと、完全文を使うこと、長い沈黙を避けること、er、you know、wellなどのつなぎ語を避けることです。

INTERACTIONS

CD114 上の絵について以下の問題に答えましょう。まずは、CDでそれぞれの問題を聞きます。問題を聞いた後に一時停止ボタンを押して、それぞれの問題に対して約1分間で回答します。回答は声に出して言いましょう。

FOLLOW-UP QUESTIONS

1. How did the pilot know there was a problem?
2. How should a pilot handle this type of landing?
3. What else should the pilot do to prepare for this emergency?

UNIT 10　　　　　　　　　　　　　　　　　　　　　　　　　　　　　　　　APPROACH

AVIATION READING

VOCABULARY　以下のキーワードを読みましょう。分からないキーワードに○をして、意味を調べましょう。

KEY WORDS
1. approach lights
2. burst into flames
3. decision
4. detect
5. final approach path
6. runway threshold
7. wreckage

PRONUNCIATION　**CD115** CDで上のキーワードを聞き、後についてリピートしましょう。何回か繰り返します。

FLUENCY　**CD116** CDで以下の文を聞き、音読しましょう。何回か音読を繰り返します。

MICROBURST

　Two of the most dangerous weather hazards that pilots face are thunderstorms and microbursts. A microburst is a very powerful downward burst of wind which spreads out in all directions after it hits the ground. Microbursts are often difficult to detect and, for pilots trying to make a final approach to land at slow speeds, the wrong decision can lead to disaster.

　On June 24th 1975, Eastern Airlines Flight 66 was making its final approach to John F. Kennedy International Airport in New York when it encountered a microburst. A very strong thunderstorm was passing over the airport at the time, and several other pilots had already reported severe windshear conditions along the final approach path to Runway 22L.

　The crew of Flight 66 received the PIREPs, but continued with their approach until the microburst caused the Boeing 727 to descend rapidly. The aircraft was 2,400 feet from the runway threshold when it struck the approach light towers. It then banked to the left, struck more approach lights, and burst into flames. Wreckage was scattered around the airport, and 113 passengers and crew died as a result of the crash.

　The NTSB investigation concluded that the ATC controllers and cockpit crew were slow to recognize that there was a dangerous microburst condition along the approach path to Runway 22L. Following this accident, low-level windshear alert systems were installed in many airports in the United States.

UNIT 11　　　　　　　　　　　　　　　　　　　　　LANDING

"GO AROUND, GO AROUND"
CROSSWINDS & WAKE TURBULENCE

VOCABULARY
以下のキーワードを読みましょう。分からないキーワードに〇をして、意味を調べましょう。

KEY WORDS

1. clear the active
2. crabbing
3. grass runway
4. gust
5. midfield
6. minimum fuel
7. obstacle
8. short field landing
9. tailwheel
10. terminal area
11. traffic flow
12. wake turbulence

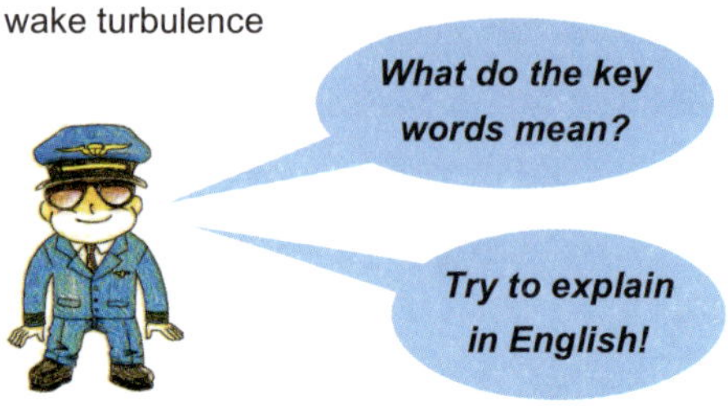

What do the key words mean?

Try to explain in English!

PRONUNCIATION　**CD117** CDで上のキーワードを聞き、後についてリピートしましょう。何回か繰り返します。

COMPREHENSION　**CD118** CDで5つのATCの会話を聞き、それぞれの問題に答えましょう。（回答は102ページ）

LISTENING TO ATC

1. (a) Describe the problem.　　　　　　　　　(b) What are the tower instructions?
2. (a) What caution does the tower issue?　　　(b) What happens after the landing?
3. (a) What is the problem with King Air 12F?　(b) How does the tower handle this problem?
4. (a) Describe the pilot's initial contact.　　　　(b) Why can't King Air 12F land on RWY 16?
5. (a) What does the tower notice?　　　　　　(b) What do you think will happen next?

FLUENCY　**CD119** CDで以下の文を聞き、音読しましょう。何回か音読を繰り返します。

DESCRIBING A SINGLE PICTURE

I can see that the aircraft is about to land on RWY 18. The ATC tower is on the left side of the runway, and I cannot see any other obstacles. The landing gear has been extended and this appears to be a normal landing. I cannot see a wind indicator, but it does not look like there is a strong crosswind. The pilot is not crabbing the airplane, so I think that the weather is good and the wind is calm. After landing, the pilot will clear the active and taxi back to the terminal area. The airport does not seem to be very busy right now.

UNIT 11   LANDING

SINGLE PICTURE

> **ADVICE** 結論を出しましょう
> 「SINGLE PICTURE」の問題では、主題を明らかにして、いくつかの話題を見つけます。そして結論を出しましょう。経験と知識を使って、次に起こる可能性がある動きや事柄を説明するようにします。
> **X** "There are many airplanes . . . Some flying . . . Some not . . ."
> **O** "There are too many airplanes in the traffic pattern, so I think this is a dangerous situation."

FLUENCY 絵の状況を描写しましょう。まず、絵の中から3つか4つの話題を見つけます（例えば、airport、traffic、runway incursion等）。次にそれぞれの話題に関して1分程度話します。目標は、3分間話すこと、完全文を使うこと、長い沈黙を避けること、er、wellなどのつなぎ語を避けることです。

INTERACTIONS **CD120** 上の絵について以下の問題に答えましょう。まずは、CDでそれぞれの問題を聞きます。問題を聞いた後に一時停止ボタンを押して、それぞれの問題に対して約1分間で回答します。回答は声に出して言いましょう。

FOLLOW-UP QUESTIONS

1. Describe the traffic flow in this picture.
2. What should the pilot turning final do if there is an aircraft on the runway just before landing?
3. How do small aircraft affect the traffic flow of large commercial airliners?

UNIT 11

LANDING

ATC DESCRIPTION

INTERACTIONS

CD121 下の絵について以下の問題に答えましょう。まずは、CDでそれぞれの問題を聞きます。問題を聞いた後に一時停止ボタンを押して、それぞれの問題に対して約1分間で回答します。回答は声に出して言いましょう。

WARM-UP QUESTIONS

1. Describe the situation in the picture.
2. What is the aircraft doing?
3. Describe the runway condition.
4. What help does the pilot need from the ground staff?

COMPREHENSION

CD122 CDでATCの会話を聞きながら3つの間違いに○をして、正しい言葉を書きましょう。
（回答は102ページ）

ATC DIALOG

CONTROLLER: OB Air 1663, Chitose Tower, cleared to land, RWY 01R, wind 030 at 20, after landing, turn left onto Taxiway Bravo 13. Caution, some snow and ice patches on the runway.

PILOT: Chitose Tower, OB Air 1663, roger, cleared to land, RWY 01R.

PILOT: OB Air 1663, can you give us the winds again?

CONTROLLER: Wind 030 at 25, peak gusts 35. Previous aircraft, a Boeing 737, reported braking action poor from midfield on.

PILOT: Roger, OB Air 1663.

PILOT: Chitose Tower, OB Air 1663, going around, RWY 01R, our approach is a little too low. Be advised, we have minimum fuel.

CONTROLLER: OB Air 1663, roger, understand your situation. Remain in right traffic, number 3, anticipate cleared to land RWY 01R.

FLUENCY

上のATCの状況を描写しましょう。1～2分話すようにします。

INTERACTIONS

CD123 上のATCの状況について、以下の問題に答えましょう。まずは、CDでそれぞれの問題を聞きます。問題を聞いた後に一時停止ボタンを押して、それぞれの問題に対して約1分間で回答します。回答は声に出して言いましょう。

FOLLOW-UP QUESTIONS

1. Do you think the pilot should have landed? Why (not)?
2. What else should the pilot have done in this situation?
3. Name some other reasons why a pilot would abort a landing and go around.

UNIT 11　　　　　　　　　　　　　　　　　　　　　　　　　　　　　　　　　　LANDING

USEFUL LANGUAGE

STRUCTURE

このエクササイズでは、「must」「has/have to」「because」を使って、とらなければならない行動とその理由を表わします。下の単語を並べ替えて文を作りましょう。1問目の例を参考にしてください。

NECESSARY ACTIONS & REASONS

1. too poor. / has to / the visibility is / The pilot / because / execute a missed approach
    → _The pilot has to execute a missed approach because the visibility is too poor._
2. have to / there is / make an emergency landing / The pilots / because / a system malfunction.
    → _____
3. because / The PF / clear the active quickly / very busy. / has to / the airport is
    → _____
4. an obstacle is / instruct the pilot / The controller / because / to go around / on the runway. / must
    → _____

PRONUNCIATION CD124
CDで上の文を聞き、後についてリピートしましょう。何回か繰り返します。（回答は102ページ）

STRUCTURE

下の絵は、さまざまな着陸状況を描写しています。下の「USEFUL STRUCTURE」とそれぞれの絵の上の単語を使って、とらなければならない行動と理由を表わす文を作りましょう。1問目の例を参考にしてください。

USEFUL STRUCTURE

NECESSARY ACTION		REASON
The pilot has to…	because	the runway is short.
The controller must…		there is a crosswind.
etc		etc

1 grass runway • houses • short field landing

2 runway incursion • cleared for T/O • go around

3 crosswind • crab • missed approach

4 tailwheel • go around • lose control

Use the structure to help you make complete sentences!

The pilot <u>has to</u> land softly <u>because</u> this is a grass runway.
She <u>has to</u> do a short field landing <u>because</u> the runway is short.
She <u>must</u> be careful <u>because</u> there are many houses around the runway.

UNIT 11 — LANDING

PICTURE SEQUENCE

Control your structure!

Remember to use past tenses for the picture sequence!

FLUENCY

1～4の順に、絵の状況を描写しましょう。
目標は、3分間話すこと、完全文を使うこと、
長い沈黙を避けること、er、you know、wellなどのつなぎ語を避けることです。

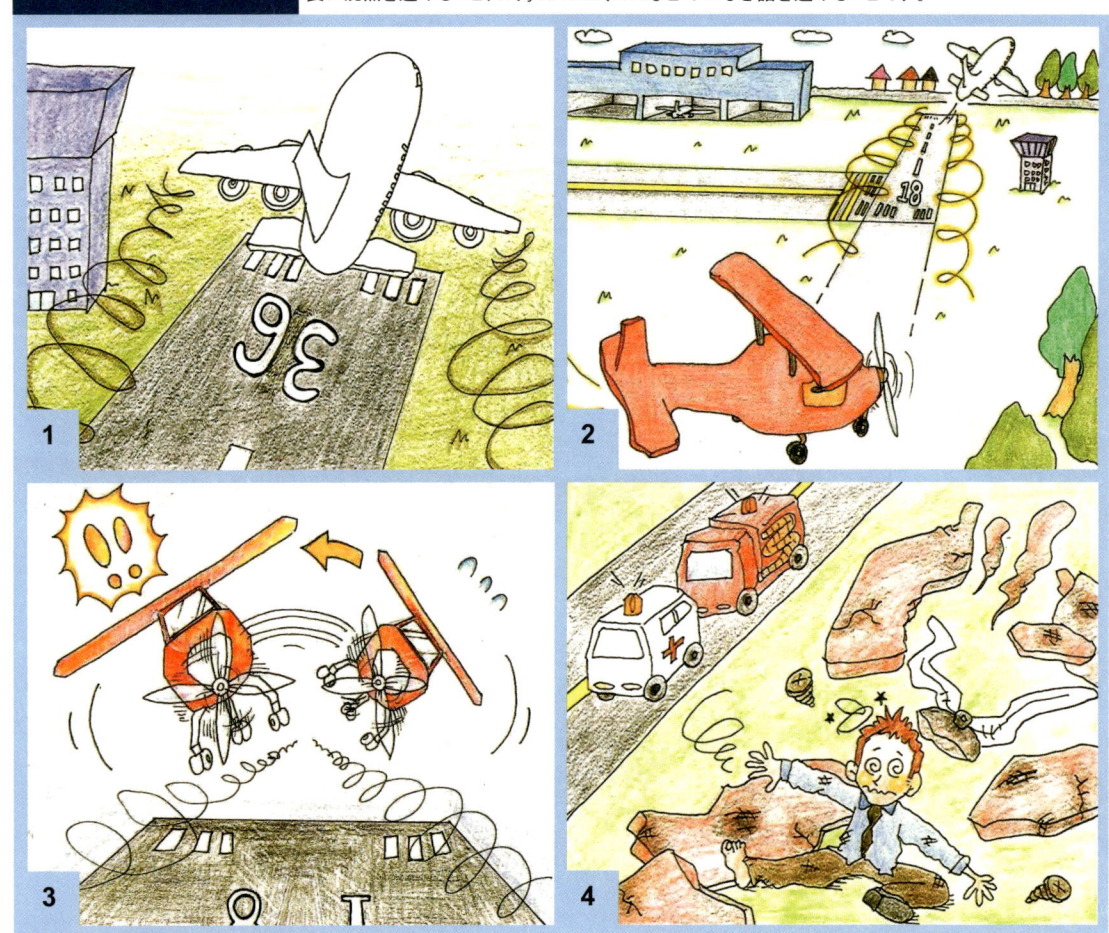

INTERACTIONS

CD125 上の絵について以下の問題に答えましょう。まずは、CDでそれぞれの問題を聞きます。問題を聞いた後に一時停止ボタンを押して、それぞれの問題に対して約1分間で回答します。回答は声に出して言いましょう。

FOLLOW-UP QUESTIONS

1. How could this accident have been prevented?
2. What is the meaning of 'wake turbulence'?
3. What advice would you give a pilot about wake turbulence?

UNIT 11   LANDING

AVIATION READING

VOCABULARY
以下のキーワードを読みましょう。分からないキーワードに○をして、意味を調べましょう。

KEY WORDS
1. belly landing
2. check item
3. checklist
4. emergency services
5. fire fighters
6. fire-fighting foam
7. gear-up landing
8. hydraulic system
9. mechanical malfunction
10. pilot error
11. substantial damage

PRONUNCIATION
CD126 CDで上のキーワードを聞き、後についてリピートしましょう。何回か繰り返します。

FLUENCY
CD127 CDで以下の文を聞き、音読しましょう。何回か音読を繰り返します。

BELLY LANDING

   A belly landing or gear-up landing happens when an aircraft lands without having its landing gear fully extended. In many cases, the cause is pilot error. Lowering the landing gear is a check item on the Before Landing checklist, but pilots who do the checklist by memory may neglect to lower the landing gear. Alternatively, pilots may forget to lower the gear if they are interrupted while performing the checklist.

   In other cases, pilots have to make belly landings because of mechanical malfunctions. On November 1st 2011, Polish Airlines Flight 16 was flying from the United States to Poland when the central hydraulic system failed. The flight crew were unable to use the alternate gear extension system to lower the landing gear, and therefore notified ATC that they would have to carry out an emergency landing.

   The Boeing 767 circled Warsaw Chopin Airport for more than one hour while emergency services prepared the airport for the landing. During this time, the pilots were using up excess fuel so that the plane could land at a lighter weight and with less risk of fire. Finally, the pilots made a successful emergency landing and fire fighters sprayed the aircraft with fire-fighting foam as the 220 passengers evacuated. There was substantial damage to the aircraft, but no injuries were reported to the passengers or crew.

UNIT 12 — AFTER LANDING

"REQUEST EMERGENCY ASSISTANCE"
OVERRUNS & OTHER MISHAPS

VOCABULARY

以下のキーワードを読みましょう。分からないキーワードに○をして、意味を調べましょう。

KEY WORDS

1. aircraft marshaller
2. airport security
3. boarding bridge
4. collision
5. deboard
6. economy class
7. evacuation order
8. first officer
9. general aviation
10. landing lights
11. skid off
12. transmission
13. wand

What do the key words mean?

Try to explain in English!

PRONUNCIATION

CD128 CDで上のキーワードを聞き、後についてリピートしましょう。何回か繰り返します。

COMPREHENSION

CD129 CDで5つのATCの会話を聞き、それぞれの問題に答えましょう。(回答は103ページ)

LISTENING TO ATC

1. (a) What seems to be the problem?
   (b) What are the tower instructions to Citation 16C?
2. (a) What are the tower instructions?
   (b) How can a pilot communicate in this situation?
3. (a) Describe the problem.
   (b) What do you think the tower's next action will be?
4. (a) What did the pilot do after landing?
   (b) What does the pilot want to do now?
5. (a) Describe the incident.
   (b) What does the pilot suggest the tower do?

FLUENCY

CD130 CDで以下の文を聞き、音読しましょう。何回か音読を繰り返します。

DESCRIBING A SINGLE PICTURE

This is an incident at the gate area. The pilot is being guided by an aircraft marshaller. The marshaller is holding wands and is showing the pilot to keep going forward. The pilot seems to be following the instructions, but he has gone too far and the left side of the cockpit has hit the boarding bridge. I think that maybe the pilot was going too fast, or the marshaller did not signal to stop in time. The passengers will not be able to get off the airplane under this condition. I think that they will deboard using the mid-section exits.

UNIT 12　　　　　　　　　　　　　　　　　　　　　　　　　　　AFTER LANDING

SINGLE PICTURE

Keep your responses smooth!

Identify the topics, then add details!

FLUENCY

絵の状況を描写しましょう。まず、絵の中から3つか4つの話題を見つけます（例えば、accident、help、evacuation等）。次にそれぞれの話題に関して1分程度話します。目標は、3分間話すこと、完全文を使うこと、長い沈黙を避けること、er、wellなどのつなぎ語を避けることです。

INTERACTIONS

CD131 上の絵について以下の問題に答えましょう。まずは、CDでそれぞれの問題を聞きます。問題を聞いた後に一時停止ボタンを押して、それぞれの問題に対して約1分間で回答します。回答は声に出して言いましょう。

FOLLOW-UP QUESTIONS

1. What do you think is going to happen next?
2. What will happen to other traffic that needs to use this airport?
3. How do you think this accident could have been prevented?

UNIT 12　　　　　　　　　　　　　　　　　　　　AFTER LANDING

ATC DESCRIPTION

INTERACTIONS

CD132 下の絵について以下の問題に答えましょう。まずは、CDでそれぞれの問題を聞きます。問題を聞いた後に一時停止ボタンを押して、それぞれの問題に対して約1分間で回答します。回答は声に出して言いましょう。

WARM-UP QUESTIONS

1. Describe the situation in the picture.
2. Why do the police officers look upset?
3. What actions should the pilot take?
4. What help does the pilot need from the ground staff?

COMPREHENSION

CD133 CDでATCの会話を聞き、下線部を埋めましょう。（回答は103ページ）

ATC DIALOG

PILOT:　　　　Tokyo Control, OB Air 1663, we have a situation in the cabin that is going to require some ___1._____. Request hold our present position on Taxiway Bravo.

CONTROLLER:　Roger, OB Air 1663, Tokyo Control, hold your position and are you declaring an emergency?

PILOT:　　　　Negative, OB Air 1663, we are not declaring an emergency. We have a male passenger, ___2._____, who has been drinking throughout the flight and also has been caught smoking in the lavatory.

CONTROLLER:　OB Air 1663, is the passenger ___3._____?

PILOT:　　　　OB Air 1663, affirm, we have the passenger in ___4._____ and he is isolated in the aft section of economy class. He is still extremely agitated and I would like him taken off board ___5._____.

CONTROLLER:　Roger, OB Air 1663, stand by for airport security. ETA is ___6._____.

FLUENCY

上のATCの状況を描写しましょう。1〜2分話すようにします。

INTERACTIONS

CD134 上のATCの状況について、以下の問題に答えましょう。まずは、CDでそれぞれの問題を聞きます。問題を聞いた後に一時停止ボタンを押して、それぞれの問題に対して約1分間で回答します。回答は声に出して言いましょう。

FOLLOW-UP QUESTIONS

1. Describe how the pilot handled this situation.
2. How else could this situation have been handled?
3. Name some other situations where a pilot would have to request police assistance.

UNIT 12 **AFTER LANDING**

USEFUL LANGUAGE

STRUCTURE

このエクササイズでは、「so」を使って、過去の状況に対する結果を表わします。下の単語を並べ替えて文を作りましょう。1問目の例を参考にしてください。

RESULTS IN THE PAST

1. it hit / was taxiing / the boarding bridge. / so / too quickly / The jet aircraft
   → _The jet aircraft was taxiing too quickly so it hit the boarding bridge._
2. to lower the landing gear / they made / so / were unable / a belly landing. / The pilots
   → _____
3. the runway / an emergency landing / made / so / The first officer / had to be closed.
   → _____
4. so / on the taxiway / There was / emergency vehicles. / the controller sent / a collision
   → _____

PRONUNCIATION

CD135 CDで上の文を聞き、後についてリピートしましょう。何回か繰り返します。（回答は103ページ）

STRUCTURE

下の絵は、さまざまな着陸後の状況を描写しています。下の「USEFUL STRUCTURE」とそれぞれの絵の上の単語を使って、過去の状況に対する結果を表わす文を作りましょう。1問目の例を参考にしてください。

USEFUL STRUCTURE

SIMPLE PAST TENSE		RESULT
There was an accident	so	the runway had to be...
The aircraft crashed		the controller sent...
etc		etc

1 approach speed • overrun • evacuation order

2 skid off • emergency vehicles • runway closed

3 landing gear • collapse • fire fighters

4 snow and ice • slippery • caution

Use past tenses for the picture sequence!

The approach speed was too high <u>so</u> the aircraft overran the runway.

There was a risk of fire <u>so</u> the captain gave the evacuation order.

The plane overran the runway <u>so</u> the controller sent emergency vehicles.

UNIT 12　　　　　　　　　　　　　　　　　　　　　　　　　　　AFTER LANDING

PICTURE SEQUENCE

ADVICE スムーズな回答を保とう

「PICTURE SEQUENCE」の問題では、話し中を通してスムーズな回答を保ちましょう。3分程度話すことが求められていますから、回答中は常に情報を流すように留意します。

X "Aircraft with fire . . . urhh . . . fire cars."　　O "Last week, there was an emergency landing because of an engine problem."

FLUENCY　1〜4の順に、絵の状況を描写しましょう。目標は、3分間話すこと、完全文を使うこと、長い沈黙を避けること、er、you know、wellなどのつなぎ語を避けることです。

INTERACTIONS　**CD136** 上の絵について以下の問題に答えましょう。まずは、CDでそれぞれの問題を聞きます。問題を聞いた後に一時停止ボタンを押して、それぞれの問題に対して約1分間で回答します。回答は声に出して言いましょう。

FOLLOW-UP QUESTIONS

1. What did the pilot do to prepare for the landing?
2. Why were the passengers only evacuating from the left side?
3. What do you think were the priorities in this emergency?

UNIT 12　　　　　　　　　　　　　　　　　　　　　　　　　　　　AFTER LANDING

AVIATION READING

VOCABULARY
以下のキーワードを読みましょう。分からないキーワードに○をして、意味を調べましょう。

KEY WORDS
1. clip
2. double-decker
3. horizontal stabilizer
4. regional jet
5. seat configuration
6. surface movement radar
7. widebody aircraft
8. wingspan
9. wingtip

ACRONYM
1. FDR

PRONUNCIATION
CD137 CDで上のキーワードを聞き、後についてリピートしましょう。何回か繰り返します。

FLUENCY
CD138 CDで以下の文を聞き、音読しましょう。何回か音読を繰り返します。

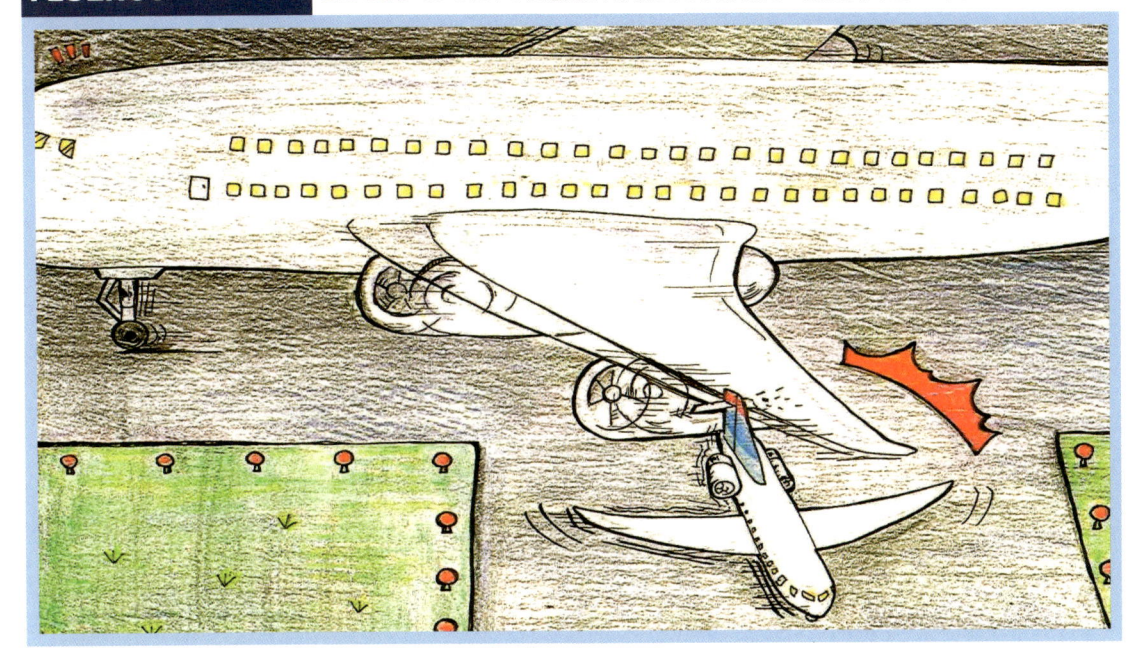

WATCH OUT!

　The Airbus A380, the world's largest passenger airliner, is easily recognizable wherever it flies. In an all-economy seat configuration, this four-engine, double-decker, widebody aircraft can carry more than 800 passengers. Its wingspan is almost 80 meters. However, the size and weight of this 'super jumbo' airplane present problems at many airports.

　One such problem happened at John F. Kennedy Airport in New York on the evening of April 11th 2011. An Air France A380 was taxiing to a runway when it clipped a Comair CRJ700 regional jet. The super jumbo's left wingtip struck the horizontal stabilizer of the Comair jet, which had just landed and was taxiing to its gate.

　There were no injuries to the passengers or crew of either aircraft. In fact, passengers on the A380 hardly noticed the incident, one passenger saying that it felt like a slight bump on the pavement. However, passengers on the smaller jet definitely noticed the contact as their aircraft was spun around almost 90 degrees.

　After the incident, NTSB investigators reviewed the flight data recorder (FDR) and cockpit voice recorder (CVR) data from both aircraft, as well as ATC tapes and surface movement radar data. In the future, as A380s become more common, more airlines and airports will have to learn how to handle these super jumbo aircraft.

UNITS 7-12

REVIEW 2

PRONUNCIATION
「B/V」の発音に注意して、以下の文を何回か音読しましょう。

B/V IN VARIOUS POSITIONS

1. The ca**b**in pro**b**lem got worse after the passenger started drinking **b**eer.
2. There was a weather ad**v**isory about **v**olcanic ash acti**v**ity.
3. The pri**v**ate pilot will go around if he is una**b**le to esta**b**lish his approach.
4. A **b**usiness class passenger is complaining about a **b**roken seat**b**elt.
5. The runway **v**isi**b**ility is o**b**scured **b**y low clouds and rain.
6. The co-pilot a**b**orted takeoff **b**ecause of **b**ird acti**v**ity on the runway.

For "V", put your top front teeth and bottom lip together, then open your mouth!

For "B", put your lips together, then open them quickly!

VOCABULARY
以下のキーワードを英語で説明してみましょう。

KEY WORDS

1. concussion
2. controlled descent
3. deboard
4. diversion
5. fatigue
6. fire fighters
7. flat tire
8. indefinitely
9. restrain
10. runway incursion
11. water ditching
12. widebody aircraft
13. wreckage

What do the key words mean?

STRUCTURE
下の単語を並べ替えて文を作りましょう。その後文を何回か音読します。(回答は 104 ページ)

SENTENCES

1. should / weather information. / contact / The pilot / for / the controller
   → _____

2. its descent / begin / The aircraft / going to / in a few minutes. / is
   → _____

3. to / The controller / the runway. / send / needs to / a tow car
   → _____

4. should / to / return / as soon as possible. / their seats / The passengers
   → _____

5. are / should execute / not seated, / If the passengers / the pilot / a missed approach.
   → _____

6. has to / because / shows minimum fuel. / the indicator / The pilot / request priority landing
   → _____

7. was activated / the captain. / the CA / A smoke alarm / so / must contact
   → _____

8. the captain. / the purser / If the passenger / to be unruly, / should contact / continues
   → _____

UNITS 7-12

REVIEW 2

PRONUNCIATION

「S/TH」の発音に注意して、以下の文を何回か音読しましょう。

For "TH", put the tip of your tongue between the top and bottom front teeth!

For "S", put the tip of your tongue behind the bottom front teeth!

S/TH IN VARIOUS POSITIONS

1. One pa**ss**enger will require immediate medical a**ss**i**s**tance.
2. I **th**ink the runway was changed due to the **s**trong cro**ss**winds.
3. The pa**ss**enger is **th**anking the pur**s**er for the good **s**ervice.
4. **S**ou**th**ea**s**terlies were reported at the de**s**tination airport.
5. The pilot will use rever**s**e **th**ru**s**t to **s**low down the airplane.
6. The 'fa**s**ten **s**eatbelt**s**' **s**ign was on when the turbulence **s**truck.

VOCABULARY

以下のキーワードを使って、下線部を埋めましょう。（回答は 104 ページ）

KEY WORDS

a. injuries
b. Kansai Airport
c. radar vectors
d. runway closure
e. smoke
f. under control

ATC DIALOG

PILOT: Tokyo Control, OB Air 1663, request emergency descent to a lower altitude due to __1._____ in the aft cabin galley. Request __2._____ to Chuubu Airport.

CONTROLLER: OB Air 1663, Tokyo Control, roger, understand your situation, unable Chuubu Airport due to __3._____. Turn right, heading 260, descend and maintain 15,000. We will give you radar vectors to __4._____. Please say the condition of your ship and passengers.

PILOT: Descending to 15,000, there are no reported __5._____ or damage at this time. The problem seems to be __6._____. Stand by for further.

FLUENCY

上のATCの状況について、以下の問題に答えましょう。

FOLLOW-UP QUESTIONS

1. Describe the situation in the ATC dialog.
2. Do you think this was an emergency situation? Why (not)?
3. Describe some situations when you would declare an emergency.

UNITS 7-12

REVIEW 2

PRONUNCIATION
「L/R」の発音に注意して、以下の文を何回か音読しましょう。

L/R IN VARIOUS POSITIONS
1. He will not file a flight plan because it is only a local flight.
2. The cabin crew went to flight operations for the briefing.
3. Hydraulic fuel is leaking from the Learjet.
4. The ferry flight was delayed because of a runway sweep.
5. The pilot received his IFR clearance from the ATC controller.
6. The passenger became unruly and violent during cruise flight.

For "L", touch the tip of your tongue to the top of your mouth!

For "R", lift the middle of your tongue but don't touch the top of your mouth!

VOCABULARY
以下のキーワードを英語で説明してみましょう。

KEY WORDS
1. belly landing
2. bruise
3. clip
4. congestion
5. defect
6. flame out
7. flight plan
8. noise abatement
9. priority landing
10. skid off
11. stack
12. traffic flow
13. unruly behavior

Use accurate sentences!

STRUCTURE
下の単語を並べ替えて文を作りましょう。その後文を何回か音読します。（回答は 105 ページ）

SENTENCES
1. weather information / needs to / at / request / The pilot / the alternate.
   → _____
2. on the runway, / an obstruction / should order / If there is / a go around. / the controller
   → _____
3. with oil. / it is contaminated / must close / The controller / because / the runway
   → _____
4. see the runway / went around. / The pilot / he / could not / so
   → _____
5. before / should / the cabin / secure / takeoff. / The cabin attendants
   → _____
6. takeoff clearance. / The pilot / for / is / contact the controller / going to
   → _____
7. will / the seatbelt sign / begin meal service / The CAs / is turned off. / after
   → _____
8. wants / his headache./ some / A passenger / for / medicine
   → _____

UNITS 7-12

REVIEW 2

VOCABULARY
以下の略称と頭字語を英語で説明してみましょう。

ABBREVIATIONS
1. CAT 1
2. PIREP
3. RWY
4. T/O
5. TWR

ACRONYMS
1. CA
2. CVR
3. ETA
4. FDR
5. FL
6. NTSB
7. PF
8. RVR

Try to explain in English!

VOCABULARY
以下のキーワードを使って、下線部を埋めましょう。(回答は105ページ)

KEY WORDS
a. chief purser
b. female passenger
c. forward lavatory
d. medical assistance
e. present position
f. Taxiway Echo

ATC DIALOG

PILOT: Tokyo Control, OB Air 1663, request hold our __1._____ on Taxiway Echo, we have a situation with one of our passengers.

CONTROLLER: Roger, OB Air 1663, Tokyo Control, hold on __2._____. Can you say the nature of your problem?

PILOT: OB Air 1663, holding on Echo, we have a __3._____, age 25-30. She is very upset and has locked herself in the __4._____.

CONTROLLER: OB Air 1663, roger, do you request __5._____?

PILOT: OB Air 1663, affirm, we would like to have medical assistance waiting at the gate, but first the __6._____ is trying to calm the passenger down and understand the situation.

FLUENCY
上のATCの状況について、以下の問題に答えましょう。

FOLLOW-UP QUESTIONS
1. Describe the situation in the ATC dialog.
2. How did the captain handle this situation?
3. What do you think happened next?

UNIT 1　　　　　　　　　　　　　　　　　　　　　　PRE-FLIGHT OPERATIONS

ANSWERS

COMPREHENSION　CD7, page 12

ATC DIALOG

CONTROLLER:　All aircraft, Haneda Airport is now closed due to a thunderstorm. Expect further information in about **30 minutes**.

PILOT:　Tokyo TWR, OB Air 1663, request estimated time of reopening.

CONTROLLER:　OB Air 1663, Tokyo TWR, the storm is moving slowly. We expect a delay of about 1 hour, possibly **shorter**.

PILOT:　Roger, request taxi back to Gate 18.

CONTROLLER:　OB Air 1663, taxi back to Gate 18 at your discretion. Use caution for strong gusting winds and **rain**.

PILOT:　Roger, OB Air 1663, Gate 18.

STRUCTURE　CD9, page 13

POSSIBLE PROBLEMS IN THE FUTURE

1. The weather may get worse later.
2. Fog may reduce visibility.
3. It could rain heavily tomorrow.
4. Lightning might strike an aircraft.
5. The thunderstorm could create microbursts.
6. Bad weather may delay the flight.
7. Severe turbulence could damage the airplane.
8. Ice on the wings might cause a loss of lift.

UNIT 2   AT THE RAMP

ANSWERS

COMPREHENSION   CD19, page 18

ATC DIALOG

CONTROLLER: OB Air 1663, Chitose Ground, cancel taxi to RWY 01R, hold your present position, acknowledge.

PILOT: Chitose Ground, OB Air 1663, holding on Taxiway Bravo. Say reason.

CONTROLLER: OB Air 1663, there is a snow removal vehicle ahead of you that is stalled. We need to call out a tow car to have it moved off the taxiway.

PILOT: Roger, our holdover time is about 15 minutes. Can you give us an estimate on the delay?

CONTROLLER: Expect a delay of about half an hour.

PILOT: Roger, request taxi back to ramp for de-icing.

CONTROLLER: OB Air 1663, approved as requested, use caution for ice and snow on the taxiway.

1. The delay is due to…
   (d) a vehicle on the taxiway
2. The estimate for the delay is…
   (d) about 30 minutes
3. The pilot decides to…
   (b) return to the ramp

STRUCTURE   CD21, page 19

EVENTS IN THE PAST

1. They de-iced the aircraft before takeoff.
2. The pilot taxied slowly because of the ice.
3. The captain decided to cancel the flight.
4. Snow continued to fall for a long time.
5. Workers removed snow from the runway.
6. Cabin crew secured the cabin before takeoff.
7. Airport staff loaded cargo onto the plane.
8. They decided to activate the anti-icing system.

UNIT 3   GROUND MOVEMENT

ANSWERS

COMPREHENSION   CD31, page 24

ATC DIALOG

PILOT: Tokyo Ground, OB Air 1663, we are **1. holding** at Taxiway Oscar. There seems to be some kind of obstruction in front of us.

CONTROLLER: OB Air 1663, Tokyo Ground, can you describe the **2. obstruction** for us?

PILOT: OB Air 1663, it looks like some kind of a box, a large rectangular box that might have fallen off a **3. baggage car**.

CONTROLLER: OB Air 1663, can you move **4. around it**?

PILOT: Negative, OB Air 1663.

CONTROLLER: Ok, OB Air 1663, hold your **5. position** until further notice. We are sending a truck to pick it up. Expect a delay of about **6. 15 minutes**.

STRUCTURE   CD33, page 25

DESCRIBING OBSTRUCTIONS

1. It seems to be a piece of baggage.
2. It looks like hydraulic fluid on the taxiway.
3. There seems to be an animal on RWY 34L.
4. It looks like it is made of plastic.
5. It seems to be about 1 metre tall.
6. It looks like there is a big piece of metal ahead.
7. There seems to be an obstruction in front of us.
8. There is some kind of container ahead of us.

UNIT 4   CLEARED FOR TAKEOFF

ANSWERS

COMPREHENSION   CD38, page 28

LISTENING TO ATC

1. (a) Because a bird sweep is in progress.   (b) They have to wait about 10 minutes.
2. (a) The tower tells King Air 12F to hold his present position.   (b) Because traffic is going around.
3. (a) King Air 12F wants to hold on the runway.   (b) Because an obstruction is at midfield.
4. (a) King Air 12F has to make an immediate departure or hold short.   (b) King Air 12F is starting the takeoff roll now.
5. (a) King Air 12F can take off from RWY 16.   (b) Because there is arriving traffic.

COMPREHENSION   CD42, page 30

ATC DIALOG

CONTROLLER: OB Air 1663, Tokyo TWR, use caution, birds in the vicinity of the airport, wind **300** at 10, RWY 34L, cleared for T/O.

PILOT 1:   Tokyo TWR, OB Air 1663, cleared for T/O, RWY 34L, starting roll.

PILOT 1:   Tokyo TWR, OB Air 1663, rejecting T/O, it seems we have hit an **obstruction** on the runway, request hold on runway.

CONTROLLER: OB Air 1663, roger, understand your situation, hold your position, RWY 34L.

PILOT 1:   OB Air 1663, holding on RWY 34L.

CONTROLLER: Park Air 7430, Tokyo TWR, go around RWY 34L, aircraft on the runway.

PILOT 2:   Roger, Park Air 7430 is going around.

PILOT 1:   Tokyo TWR, OB Air 1663, we are going to need a tow. It seems we have a **tire burst**.

CONTROLLER: Roger, OB Air 1663, standby for your request.

STRUCTURE   CD44, page 31

NECESSARY ACTIONS

1. The pilots have to hold their position.
2. Airport staff must remove the birds.
3. The captain has to request tow assistance.
4. The flight must be canceled.
5. They have to wait until the runway is active.
6. The PF must continue straight ahead.
7. The passengers have to get off the airplane.
8. The runway has to be closed for 30 minutes.

UNIT 5 — TAKEOFF & CLIMB

ANSWERS

COMPREHENSION CD49, page 34

LISTENING TO ATC

1. (a) They have to shut down the left engine.
   (b) He wants to return to the airport.
2. (a) He must climb then turn right after 800 feet.
   (b) He has to look out for birds.
3. (a) He wants to return to the ramp because a passenger needs medical assistance.
   (b) The passenger is an older male who may have had a heart attack.
4. (a) An obstruction is on the taxiway. It may be a dead animal.
   (b) He may instruct the pilot to hold his position, then send a car to remove the obstruction.
5. (a) The instructions are to hold position and give way to a Learjet, then continue taxiing to RWY 34.
   (b) There might be ice on the taxiway. The taxiway condition is reported slippery.

COMPREHENSION CD53, page 36

ATC DIALOG

CONTROLLER: OB Air 1663, wind 030 at 10, RWY 34L, cleared for T/O, climb and maintain 8,000, turn right heading 130, contact Tokyo Departure on 126.0

PILOT: Cleared for T/O, RWY 34L, 8,000, HDG 130, 126.6, OB Air 1663.

CONTROLLER: Negative. That's 126.0 for Tokyo Departure.

PILOT: Roger, 126.0, good day.

PILOT: Tokyo Departure, OB Air 1663, climbing through 1,500, HDG 130, request air turnback, we have a problem with one of our controls.

CONTROLLER: OB Air 1663, Tokyo Departure, climb and maintain 3,000, continue turning right heading 160, we'll set you up for RWY 34L and are you declaring an emergency?

PILOT: Negative on the emergency. It seems we have a mechanical problem with our flap. It is stuck and we cannot retract it. Request a full stop on 34L.

CONTROLLER: Roger, OB Air 1663, you are cleared to land, RWY 34L.

PILOT: Thank you, cleared to land, RWY 34L, OB Air 1663.

1. OB Air 1663 is instructed to… **(a) climb to 8,000**
2. OB Air 1663 wants to… **(b) return to the airport**
3. The problem seems to be… **(d) a flap in the wrong position**

STRUCTURE CD55, page 37

EVENTS HAPPENING NOW

1. A red warning light is flashing.
2. Birds are flying around the airport.
3. The airplane is experiencing engine problems.
4. A dead animal is obstructing traffic.
5. The pilots are holding on the taxiway.
6. The captain is shutting down the engine.
7. The plane is returning to the departure airport.
8. A Learjet is approaching on the taxiway.

UNIT 6　　　　　　　　　　　　　　　　　　　　　　　　　　　　　　　　CLIMB

ANSWERS

COMPREHENSION   CD60, page 40

LISTENING TO ATC

1. (a) He is reporting a bird strike.
   (b) He thinks a bird might have hit the wings.
2. (a) He encountered rough air between 4,500 and 6,000 feet.
   (b) He wants to change heading from 240 to 300 after the aircraft reaches 10,000 feet.
3. (a) He is worried about an obstruction at the end of the runway.
   (b) He says that they will do a visual check of the engines and wings.
4. (a) A passenger wants to return to the gate because he cannot find his passport.
   (b) They will probably search for the passport. If they cannot find it, they may return to the gate.
5. (a) He must return to the ramp to de-ice again.
   (b) It is the time after de-icing that an aircraft can wait before takeoff.

COMPREHENSION   CD64, page 42

ATC DIALOG

PILOT:　　　　　Fukuoka Control, OB Air 1663, climbing out of 8,000 for **1. 10,000**. We have a PIREP.

CONTROLLER:　　OB Air 1663, Fukuoka Control, **2. radar contact**, go ahead with your PIREP.

PILOT:　　　　　OB Air 1663, we encountered possible **3. volcanic ash cloud** between 5,000 and 8,000, 20 miles southeast of Mt. Sakura. Request heading change to avoid further encounter.

CONTROLLER:　　OB Air 1663, roger, turn right, **4. HDG 220**, climb and maintain 10,000. Say condition of your aircraft.

PILOT:　　　　　OB Air 1663, right turn, HDG 220, climb and maintain 10,000. Our windshield is slightly scratched and obscuring our **5. vision**, but all engine instruments indicate normal operation. We will continue on to our destination airport.

CONTROLLER:　　Roger, OB Air 1663, thank you for your **6. report**.

STRUCTURE   CD66, page 43

INTERRUPTED ACTIONS IN THE PAST

1. The aircraft was levelling off when the cockpit heat failed.
2. The plane was climbing to FL 270 when it encountered severe turbulence.
3. The cabin attendants were serving drinks when the bird strike happened.
4. The pilots were climbing through 10,000 feet when a cockpit warning sounded.

REVIEW 1

ANSWERS

STRUCTURE   Page 46

SENTENCES

1. The controller has to send a tow car to the runway.
2. Many birds are flying near the end of the runway.
3. The captain climbed out of the icing conditions.
4. There seems to be a suitcase on the taxiway.
5. Oil is leaking from the airplane onto Taxiway Bravo.
6. The runways were very wet and slippery yesterday.
7. The airplane was leveling off when lightning struck the wing.
8. Volcanic ash might cause the engines to malfunction during flight.

VOCABULARY   Page 47

ATC DIALOG

CONTROLLER: All aircraft, **1. Chitose Airport** is now closed due to a **2. snowstorm**. Expect further information in about 30 minutes.

PILOT: Chitose TWR, OB Air 1663, request estimated time of **3. re-opening**.

CONTROLLER: OB Air 1663, Chitose TWR, the storm is moving quickly, but we will need to remove snow from **4. all areas** of the airport. Time of re-opening is unknown.

PILOT: Roger, request taxi back to **5. Gate 22**.

CONTROLLER: OB Air 1663, taxi back to Gate 22 at your discretion. Taxi with extreme caution due to blowing snow and **6. reduced visibility**.

PILOT: Roger, OB Air 1663, Gate 22.

REVIEW 1

ANSWERS

STRUCTURE — Page 48

SENTENCES

1. Snow is covering the taxiways and runways at Haneda Airport.
2. There is some kind of obstruction at the ramp area.
3. The pilot did not receive his clearance before taking off.
4. The aircraft was taxiing to the runway when the fog began to lift.
5. The pilot is making an emergency landing due to a sick passenger.
6. The controller may change runways if the wind direction changes.
7. The passengers must remain seated during takeoffs and landings.
8. The pilot was holding on the runway when the earthquake began.

VOCABULARY — Page 49

ATC DIALOG

CONTROLLER: OB Air 1663, Tokyo TWR, use caution, **1. microbursts** reported in the vicinity of the airport, wind 150 at 16, RWY 16RL, cleared for T/O.

PILOT: Tokyo TWR, OB Air 1663, cleared for T/O, RWY 16R, starting **2. takeoff roll**.

PILOT: Tokyo TWR, OB Air 1663, rejecting T/O, we seem to have a problem with our **3. main landing gear**, we are holding on the runway.

CONTROLLER: OB Air 1663, roger, understand your situation, hold your position, we now see some kind of **4. fluid** leaking from your aircraft.

PILOT: OB Air 1663, we are getting an indication from our hydraulics. Request towback to **5. ramp area**.

CONTROLLER: Roger, OB Air 1663, we'll send a **6. tow truck** to the runway.

UNIT 7   CRUISE

ANSWERS

COMPREHENSION CD71, page 50

B/V IN INITIAL POSITION

1. **b**an
2. **v**at
3. **v**eer
4. **b**ent
5. **v**ery
6. **v**est
7. **v**et
8. **b**oat
9. **v**olt
10. **b**owl

COMPREHENSION CD76, page 52

ATC DIALOG

PILOT: San Francisco, OB Air 1663, we have a problem with one of our passengers, white male, approximately 35-40 yrs old. He is traveling alone and has been drinking during the entire flight. Our cabin crew has reported he has become violent and I have ordered him to be restrained.

CONTROLLER: OB Air 1663, understand your situation, say your intentions.

PILOT: OB Air 1663, the passenger is now restrained and has calmed down. Our ETA to Honolulu International Airport is 2 hours and 35 minutes. Request police assistance. Also, one of our cabin crew has been injured. She will require medical assistance on arrival.

CONTROLLER: OB Air 1663, roger, we'll relay this information to HCF Approach. Do you want us to relay a request for a closer gate?

PILOT: Negative, we'll taxi to our original gate.

CONTROLLER: Roger, OB Air 1663, contact us if there are any changes in the situation.

PILOT: Roger, over, OB Air 1663.

(Note: The controller in this dialog is SFO AIRINC.)

1. The passenger is…
   (c) drunk
2. The pilot wants to…
   (d) continue to the destination
3. The pilot does not want…
   (d) to stop on the runway

STRUCTURE CD78, page 53

SOLVING PROBLEMS

1. The purser should make a doctor call.
2. The captain should give the evacuation order.
3. Passengers should fasten their seatbelts.
4. The drunk man should not drink more alcohol.
5. The crew should restrain the unruly passenger.
6. The pilots should request police assistance.
7. The CAs should not move the pregnant woman.
8. The CA should bring some medicine.

UNIT 8 — EMERGENCY

ANSWERS

COMPREHENSION CD83, page 56

S/TH IN INITIAL POSITION

1. **th**ank
2. **s**aw
3. **s**ick
4. **th**igh
5. **th**in
6. **s**ing
7. **th**ink
8. **th**umb
9. **s**ong
10. **th**ought

COMPREHENSION CD88, page 58

ATC DIALOG

PILOT: Tokyo Control, OB Air 1663, we are making an emergency descent to a lower altitude due to **smoke** in the cabin. The oxygen masks have been deployed. Request radar vectors to the nearest airport.

CONTROLLER: OB Air 1663, Tokyo Control, roger, understand your situation, turn right, heading **180**, descend and maintain 10,000. We will give you radar vectors to Chuubu Airport. Please say the condition of your ship and passengers.

PILOT: Descending to 10,000, all passengers are accounted for, **four** injuries reported. We do not know the extent of the damage yet, but there seems to be some problem near the overhead baggage compartment in the forward section of the cabin.

STRUCTURE CD90, page 59

ACTIONS IN THE FUTURE

1. The captain will request police assistance.
2. The pilot is going to make a controlled descent.
3. The crew will restrain the unruly passenger.
4. The purser is going to check the lavatory.
5. The pilots are going to request a priority landing.
6. The cabin crew will put on oxygen masks.
7. The captain will give the evacuation order.
8. The CAs are going to use the fire extinguishers.

UNIT 9   HOLDING

ANSWERS

COMPREHENSION CD95, page 62

L/R IN INITIAL POSITION

1. lack
2. lake
3. ramp
4. land
5. rain
6. rate
7. lead
8. leak
9. lift
10. right
11. load
12. wrong
13. low

COMPREHENSION CD100, page 64

ATC DIALOG

CONTROLLER: All aircraft, Haneda Airport is now closed due to a major **1. earthquake**. We need to do a runway sweep of all runways before **2. re-opening**. Stand by for further. BREAK BREAK. OB Air 1663, Tokyo Tower, go around, turn right heading 100, climb and maintain 7,000, hold over CHIBA.

PILOT: OB Air 1663, roger, 7,000, right heading 100, be advised, we can only hold 60 minutes, request condition of Narita Airport for **3. possible diversion**.

CONTROLLER: OB Air 1663, Narita Airport is now closed. Narita is on **4. standby power** and damage reported to the tower facility. Re-open **5. schedule** is unknown at this time.

PILOT: OB Air 1663, roger, we'll hold until we get more information from the company.

CONTROLLER: Understand, OB Air 1663, we'll keep you updated on the situation, but it appears to have been a pretty big earthquake. A **6. tsunami warning** has been issued at Haneda Airport.

STRUCTURE CD102, page 65

WANTS & NEEDS

1. The pilot wants to get more information.
2. The CA needs to check the emergency exit.
3. The captain wants to request police assistance.
4. The pilots need to wait for clearance.
5. The controller wants to know the ETA.
6. They want to solve the problem quickly.
7. The passenger wants to move to another seat.
8. The pilots need to divert to the alternate airport.

UNIT 10 — APPROACH

ANSWERS

COMPREHENSION CD107, page 68

LISTENING TO ATC

1. (a) The wind is 190 at 18 knots, with gusts at 24 knots.
   (b) The pilot should go around if his approach is not stable.
2. (a) There is a landing gear indication on the right landing gear.
   (b) They want the tower to make a visual check.
3. (a) Aircraft should fly over St. Joseph's Hospital at or above 3,500 feet.
   (b) It is a procedure to minimize the impact of aircraft noise on the areas around airports.
4. (a) The instructions are to go around RWY 34 and remain in left closed traffic.
   (b) There is a runway incursion. A Cessna needs a tow because of a flat tire.
5. (a) They are going to execute a missed approach.
   (b) Because they are not CAT 1 approved.

COMPREHENSION CD111, page 70

ATC DIALOG

CONTROLLER: OB Air 1663, Chitose Tower, cleared to land, RWY 01R, wind variable at 3 knots, visibility 1000, fog.
PILOT: Cleared to land, RWY 01R, OB Air 1663.
PILOT: Chitose Tower, OB Air 1663, we are having trouble seeing the runway, say again the RVR.
CONTROLLER: OB Air 1663, RVR dropping down to 500 meters.
PILOT: Chitose Tower, OB Air 1663, we are going around, we can't make out the runway.

1. OB Air 1663 is first instructed to…
   (a) land on RWY 01R
2. The pilot is worried about…
   (c) runway visibility
3. The pilot finally decides to…
   (d) execute a missed approach

STRUCTURE CD113, page 71

CONDITIONAL ADVICE

1. If there are birds on the runway, the pilot should go around.
2. If the approach speed is too fast, the PF should reduce speed.
3. If there is windshear on approach, the pilot should not try to land.
4. If the approach is too low, the pilot should increase speed and pitch.

UNIT 11 — LANDING

ANSWERS

COMPREHENSION CD118, page 74

LISTENING TO ATC

1. (a) There is a sick passenger with chest pains on board the aircraft.
   (b) The instructions are to land on RWY 34, then turn left on the high speed and taxi to gate 12.

2. (a) RWY 16 is contaminated due to ice and the braking action is poor.
   (b) The aircraft is unable to taxi because the brakes have overheated.

3. (a) The aircraft has landed past the taxiway.
   (b) The tower instructs the pilot to taxi back to Taxiway Bravo and clear the active.

4. (a) He is 15 miles south of the airport. He wants to land on RWY 16 and has Information Sierra.
   (b) RWY 16 is closed due to bird scaring.

5. (a) A panel seems to have fallen from Citation 16C.
   (b) If he thinks it is safe to do so, the pilot may continue taxiing to the ramp area.

COMPREHENSION CD122, page 76

ATC DIALOG

CONTROLLER:	OB Air 1663, Chitose Tower, cleared to land, RWY 01R, wind 030 at 20, after landing, turn **right** onto Taxiway Bravo 13. Caution, some snow and ice patches on the runway.
PILOT:	Chitose Tower, OB Air 1663, roger, cleared to land, RWY 01R.
PILOT:	OB Air 1663, can you give us the winds again?
CONTROLLER:	Wind 030 at 25, peak gusts 35. Previous aircraft, a Boeing 737, reported braking action **good** from midfield on.
PILOT:	Roger, OB Air 1663.
PILOT:	Chitose Tower, OB Air 1663, going around, RWY 01R, our approach is a little too **high**. Be advised, we have minimum fuel.
CONTROLLER:	OB Air 1663, roger, understand your situation. Remain in right traffic, number 3, anticipate cleared to land RWY 01R.

STRUCTURE CD124, page 77

NECESSARY ACTIONS & REASONS

1. The pilot has to execute a missed approach because the visibility is too poor.
2. The pilots have to make an emergency landing because there is a system malfunction.
3. The PF has to clear the active quickly because the airport is very busy.
4. The controller must instruct the pilot to go around because an obstacle is on the runway.

UNIT 12 AFTER LANDING

ANSWERS

COMPREHENSION CD129, page 80

LISTENING TO ATC

1. (a) There is a runway incursion on RWY 16.
   (b) The instructions are to go around and remain in left closed traffic.

2. (a) The instructions are to give way to the Citation, then taxi to the ramp.
   (b) The pilot can use the landing lights or try switching to another radio.

3. (a) There is some fluid at midfield on the runway.
   (b) I think the controller will first close the runway, then send someone to clean up the fluid.

4. (a) He shut down the left engine.
   (b) He wants tow assistance to the ramp area.

5. (a) The aircraft skidded on the taxiway and the right wingtip struck the chain fence.
   (b) He suggests that the tower send someone to check the fence.

COMPREHENSION CD133, page 82

ATC DIALOG

PILOT: Tokyo Control, OB Air 1663, we have a situation in the cabin that is going to require some **1. police assistance**. Request hold our present position on Taxiway Bravo.

CONTROLLER: Roger, OB Air 1663, Tokyo Control, hold your position and are you declaring an emergency?

PILOT: Negative, OB Air 1663, we are not declaring an emergency. We have a male passenger, **2. early twenties**, who has been drinking throughout the flight and also has been caught smoking in the lavatory.

CONTROLLER: OB Air 1663, is the passenger **3. restrained**?

PILOT: OB Air 1663, affirm, we have the passenger in **4. plastic loops** and he is isolated in the aft section of economy class. He is still extremely agitated and I would like him taken off board **5. immediately**.

CONTROLLER: Roger, OB Air 1663, stand by for airport security. ETA is **6. 10 minutes**.

STRUCTURE CD135, page 83

RESULTS IN THE PAST

1. The jet aircraft was taxiing too quickly so it hit the boarding bridge.
2. The pilots were unable to lower the landing gear so they made a belly landing.
3. The first officer made an emergency landing so the runway had to be closed.
4. There was a collision on the taxiway so the controller sent emergency vehicles.

REVIEW 2

ANSWERS

STRUCTURE   Page 86

SENTENCES

1. The pilot should contact the controller for weather information.
2. The aircraft is going to begin its descent in a few minutes.
3. The controller needs to send a tow car to the runway.
4. The passengers should return to their seats as soon as possible.
5. If the passengers are not seated, the pilot should execute a missed approach.
6. The pilot has to request priority landing because the indicator shows minimum fuel.
7. A smoke alarm was activated so the CA must contact the captain.
8. If the passenger continues to be unruly, the purser should contact the captain.

VOCABULARY   Page 87

ATC DIALOG

PILOT: Tokyo Control, OB Air 1663, request emergency descent to a lower altitude due to **1. smoke** in the aft cabin galley. Request **2. radar vectors** to Chuubu Airport.

CONTROLLER: OB Air 1663, Tokyo Control, roger, understand your situation, unable Chuubu Airport due to **3. runway closure**. Turn right, heading 260, descend and maintain 15,000. We will give you radar vectors to **4. Kansai Airport**. Please say the condition of your ship and passengers.

PILOT: Descending to 15,000, there are no reported **5. injuries** or damage at this time. The problem seems to be **6. under control**. Stand by for further.

REVIEW 2

ANSWERS

STRUCTURE — Page 88

SENTENCES

1. The pilot needs to request weather information at the alternate.
2. If there is an obstruction on the runway, the controller should order a go around.
3. The controller must close the runway because it is contaminated with oil.
4. The pilot could not see the runway so he went around.
5. The cabin attendants should secure the cabin before takeoff.
6. The pilot is going to contact the controller for takeoff clearance.
7. The CAs will begin meal service after the seatbelt sign is turned off.
8. A passenger wants some medicine for his headache.

VOCABULARY — Page 89

ATC DIALOG

PILOT: Tokyo Control, OB Air 1663, request hold our **1. present position** on Taxiway Echo, we have a situation with one of our passengers.

CONTROLLER: Roger, OB Air 1663, Tokyo Control, hold on **2. Taxiway Echo**. Can you say the nature of your problem?

PILOT: OB Air 1663, holding on Echo, we have a **3. female passenger**, age 25-30. She is very upset and has locked herself in the **4. forward lavatory**.

CONTROLLER: OB Air 1663, roger, do you request **5. medical assistance**?

PILOT: OB Air 1663, affirm, we would like to have medical assistance waiting at the gate, but first the **6. chief purser** is trying to calm the passenger down and understand the situation.

SELF-EVALUATION

This page helps check your English proficiency in the 6 areas of evaluation. Read each sentence below and circle "True" or "False".

After you have finished, check the sentences circled "False". Which areas of evaluation are they? Practice these areas more.

To achieve Level 4 on the English language proficiency test, you should be able to answer "True" for all the sentences below. Do your best!

What are your weak areas?

PRONUNCIATION

• My pronunciation is clear and accurate.	TRUE	FALSE
• My pronunciation is usually easy to understand.	TRUE	FALSE

STRUCTURE

• I speak in complete sentences.	TRUE	FALSE
• I control my structure and tenses.	TRUE	FALSE
• I do not make mistakes that change the meaning of my sentences.	TRUE	FALSE

VOCABULARY

• I know a wide range of aviation vocabulary.	TRUE	FALSE
• I make accurate word choices when needed.	TRUE	FALSE

FLUENCY

• I can speak at length on aviation topics.	TRUE	FALSE
• My speech is smooth and is not broken by fillers or pauses.	TRUE	FALSE
• I connect my ideas with discourse markers (eg: "so", "because", "therefore").	TRUE	FALSE

COMPREHENSION

• I usually understand what other speakers are saying.	TRUE	FALSE
• I always check and confirm misunderstandings.	TRUE	FALSE

INTERACTIONS

• I can maintain communication exchanges.	TRUE	FALSE
• I offer enough information.	TRUE	FALSE
• My responses are timely and accurate.	TRUE	FALSE

Simon Cookson has taught at J. F. Oberlin University since 2001. He is currently an Associate Professor in the Aviation Management Department.

He has a Ph. D. in Sociolinguistics from International Christian University, a Master's degree in Aerospace Systems Engineering from the University of Southampton in the UK, and a Master's degree in Teaching English to Speakers of Other Languages (TESOL) from Aston University in the UK.

His main research interests are corpus linguistic approaches to the language training of flight crew and airline accidents that involve language factors. He has published a number of research papers about accidents cited by ICAO to justify its English language proficiency program.

サイモン・クックソン

2001年より桜美林大学教員。現在は同大学アビエーションマネジメント学類准教授。

国際基督教大学社会言語学博士号、英国サウサンプトン大学院、航空宇宙システムエンジニアリング修士号、および英国アストン大学院、英語教授法修士号を持つ。

主な研究分野は、コーパス言語学を取り入れたフライトクルー教育と、言語要因が関わる航空事故。ICAO が英語能力プログラム導入の正当化のために引証している航空事故ケースに関し、多数の研究論文を発表している。

Michael Kelly worked for 23 years in the Flight Training Department at Japan Airlines managing the aviation and ATC English programs for pilot trainees. He is now an Assistant Professor in the Aviation Management Department at J. F. Oberlin University.

He holds a JCAB PPL license and an Aeronautical Radio Operator license, and is approved to evaluate foreign pilots for the Aeronautical Radio Operator test.

His main research interest is the use of phraseology versus plain English in ATC-pilot communication.

マイケル・ケリー

23年間、日本航空のフライトトレーニングデパートメント勤務。パイロット訓練生の航空英語とATC英語のプログラム管理を行った。現在は桜美林大学アビエーションマネジメント学類専任講師。

JCABの自家用操縦士ライセンス、および航空無線通信士資格を持つ。また、外国人受験者の航空無線通信士試験の認定試験官である。

主な研究分野は、phraseologyと日常英語を使用したパイロットと管制官のコミュニケーション。

READY FOR DEPARTURE!
パイロットのための
ICAO 航空英語能力試験教本

定価はカバーに表示してあります

2012 年 5 月 18 日　初版発行
2025 年 6 月 18 日　6 版発行

著　者　サイモン・クックソン，マイケル・ケリー
発行者　小川　啓人
印　刷　倉敷印刷株式会社
製　本　東京美術紙工協業組合

発行所　株式会社 成山堂書店

〒160-0012　東京都新宿区南元町 4 番 51 成山堂ビル
TEL：03(3357)5861　FAX：03(3357)5867
URL　https://www.seizando.co.jp
落丁・乱丁本はお取り換えいたしますので、小社営業チーム宛にお送りください。

Ⓒ 2012　Simon Cookson, Michael Kelly
Printed in Japan　　　　　　　ISBN978-4-425-86211-5

附属 CD-ROM のご利用について

■ファイル形式は mp3 です。対応する機器（PC、デジタルオーディオプレーヤー等）でご利用ください。音楽 CD 専用機ではご利用いただけない場合がございます。その場合は、mp3 ファイルをオーディオファイルに変換するなどしてご利用ください。

■CD-ROM の複製は個人で利用する範囲とします。著者または出版社の許可なく、他者に複製を供与する、またはインターネット上に公開する等の無断複製・頒布行為を禁じます。